GW01606971

ORNAMENTAL TURNERY

ORNAMENTAL TURNERY

A Practical and Historical Approach to a Centuries-Old Craft

FRANK M. KNOX

Foreword by Wilfred John Osborne

PRENTICE HALL PRESS • NEW YORK

The extract on pps. 14–15 is from *Miss Willmott of Warley Place: Her Life and Her Gardens* by Dr. Audrey le Lievre and is reprinted with permission of the publisher, Faber and Faber Limited.

Published by Prentice Hall Press
A Division of Simon & Schuster, Inc.
Gulf + Western Building
One Gulf + Western Plaza
New York, NY 10023

PRENTICE HALL PRESS is a trademark of Simon & Schuster, Inc.

Library of Congress Cataloging-in-Publication Data

Knox, Frank M.
Ornamental turnery.

Includes index.
1. Turning—Amateurs' manuals. I. Title.
TT203.K55 1986 684'.083 86-3187
ISBN 0-671-61369-3

Designed by Jack Meserole/Publishing Synthesis Ltd.

Manufactured in the United States of America
10 9 8 7 6 5 4 3 2 1
First Edition

To Dorothy,
for forty-nine wonderful years

ACKNOWLEDGMENTS

For an amateur craftsman to write a book on a subject that has not seen a new treatise in more than one hundred years requires not only a great amount of courage but, more important, a great deal of help from others who have maintained an interest in the subject. Such help has been generously provided by members of the Society of Ornamental Turners, both directly on request and through articles written by society members for publication in their semiannual bulletins. I would like to acknowledge the outstanding assistance provided by the following individuals and organizations:

Wilfrid John Osborne, of Stratford-on-Avon, England. Years ago, when I first discovered that ornamental turnery existed, I called for help in obtaining a lathe. Mr. Osborne took it upon himself to search for, discover, and take possession of such a lathe, notifying me of his find and arranging for its transportation from England to New York, where I have actively used it for more than twenty years. Moreover, because there are few active ornamental turners in the United States, he has become a valued correspondent on troublesome questions, searching further for missing or additional material. I owe Wilfrid John Osborne an immeasurable debt of gratitude for getting me started on a most satisfying creative occupation. He gave freely of his time in reviewing the manuscript for this book and offered valuable constructive criticism, for which he was eminently qualified, having been editor of the *Bulletin of the Society of Ornamental Turners* for many years.

Warren Greene Ogden, Jr., of North Andover, Massachusetts. Mr. Ogden is a top-level engineer and precision instrument maker who has made a life study of the Holtzapffel family and their achievements in designing and making ornamental turning lathes. Much of the material in this book on the history of ornamental turnery, particularly that pertaining to the lathe and the Holtzapffel family, came from Mr. Ogden's extensive research files and from his association with the British Science Museum. As a master toolmaker, he was of invaluable assistance in making and repairing broken or missing parts for my lathe, an inevitable problem with a 130-year-old lathe. Again, the manuscript for this book was carefully reviewed and corrected by Mr. Ogden.

Delphin Broussailles of New York, New York. An internationally known artist and craftsman, Mr. Broussailles uses a superb Holtzapffel lathe to form and ornament ivory and later further ornaments his work using a centuries-old Etruscan process known as *gold granulation.* Not more than an estimated half-dozen persons in the world today use this technique, which consists of taking granules of gold as fine as 1/8000 inch, placing them in patterns, and affixing them to a 22-carat-gold base. A retired associate professor of the history of fine arts, he has greatly assisted in relating the craft of ornamental turnery to the field of art as a whole, thus clarifying its relation to the decorative arts.

Richard Miller of Westfield, New Jersey. An ornamental turner and owner of a fine Evans lathe, Mr. Miller gave unstinting assistance by researching ornamental turnery in his extensive collection of antique books on the subject. He was instrumental in arranging, financing, and actually manning the booths and demonstrating turning techniques for several shows and exhibitions of ornamental turnery.

Herbert Jaffee of New York, New York. A master toolmaker, Mr. Jaffee was a dependable helper when my lathe developed mechanical problems that were beyond my ability to fix. His on-the-spot analyses of operating problems helped me over many rough spots. His technical review of the manuscript for this book was also of great assistance.

Walter Balliet of Collingswood, New Jersey. A retired master toolmaker with more than forty years of experience, Mr. Balliet has built a complete ornamental lathe in the manner of Holtzapffel. His aid in the technical aspects of the lathe and its use is greatly appreciated.

The Society of Ornamental Turners. Too numerous to list individually are the members of the Society of Ornamental Turners who have contributed articles to the society's bulletin. It consequently is a veritable gold mine of information on ornamental turnery and its related subjects. I have collected a complete set of the bulletins and spent many hours reviewing them for material for this book.

Finally, this book would probably never have seen the light of day in other than my own highly questionable handwriting had it not been for my devoted wife, Dorothy, who took time from her multitudinous club duties to type and correct the several drafts and final version of the manuscript.

I am indebted to all the people who helped me to develop the concept for and complete the enjoyable task of writing this book on one aspect of craftsmanship in the decorative arts.

CONTENTS

FOREWORD

More than a century has elapsed since the magnum opus *Turning and Mechanical Manipulation* was published by John Jacob Holtzapffel and his family. A most comprehensive study, it remained out of print until Robert Johnson and Frank Knox consulted with Dover Publications on reprinting the work. Holtzapffel's book is a masterly exposition of the craft but was written by an engineer in a somewhat tortuous style, with eccentric punctuation. The aforementioned gentlemen were members of the Society of Ornamental Turners (S.O.T), which was formed in 1948 to revive interest in this delightful craft. The Dover reprint has been remarkably successful despite its flaws and despite the fact that ornamental lathes are no longer made and are therefore rare.

Stirred by the international sales of the Dover reprint, Frank Knox decided that an entirely new book on ornamental turnery would appeal to many enthusiasts like himself who had discovered the infinite possibilities of this fascinating craft. That it was practiced by emperors, kings, and princes is of little account save that ornamental turnery gained a certain snob appeal. Snob appeal should not be scorned, for it can play a part in increasing and broadening aesthetic appreciation—just think of Fabergé and his marvelous work with jeweled eggs.

But why should anyone be interested in ornamental turnery? The reason is simple. Societies such as the S.O.T exist to clarify and strengthen the role of the craftsman in a world where cybernetics and technology reign supreme. Enthusiasts like the author realize that the world is losing respect for manual skills. To millions the word *craftsman* is used to distinguish the trained journeyman from the semiskilled. It has no reference to aesthetic values but refers rather to union struggles over pay differentials.

The craftsman is not just an assembly-line worker, using manual skills to build the same part of the same product over and over again. The creative craftsman is concerned with making one complete article. He is resonsible for the whole process, from designing to finishing the piece. He enjoys this responsibility. The ornamental turner is responsible for all aspects of the work he creates, from the tools he uses, to the shape and finish of his product. He uses his intellect as well as his manual skill to create an article of beauty. He decides upon the best use of his materials and the particular requirements of his design to achieve the effect he wants, and then he uses his manual skill to make sure that the finished object is as technically correct as it can possibly be.

Ornamental turning lathes are not made commercially today but this fact should not prevent anyone sufficiently interested in the craft from practicing it. Metal-working lathes can be adapted for ornamental turnery, and they are capable of performing many operations. These adaptations are well within the scope of engineers, and several members of the S.O.T have modified equipment that is now suitable for the finest work. In addition, with a modern slide rest, a wood-turning lathe with end-turning attachments is capable of producing fine work.

This book is certain to appeal to a wide audience, as it discusses all the problems that a newcomer to ornamental turnery confronts. It provides a detailed analysis of the craft and of the use of the various pieces of equipment peculiar to the ornamental lathe. It is impossible in one lifetime to exhaust the possibilities of ornamental turnery, with its limitless scope for the artist, the designer, and the technician.

The lyf so short, the craft so long to lerne
Th'assay so hard, so sharp the conquering.

So wrote Chaucer six centuries ago. The truth of his words is demonstrated by Frank Knox with the artifacts he illustrates and the fund of knowledge he imparts in this book. It will be of inestimable value to all interested in craftsmanship.

—WILFRID JOHN OSBORNE
Former editor of the *Bulletin of the Society of Ornamental Turners*
Stratford-on-Avon, England

PREFACE

This book was written by an amateur for other amateurs. It deals with a very popular subject, wood turning, but with a very little known and exotic branch of that craft, ornamental turnery. Although I have had my ornamental lathe for twenty years and had studied cabinetmaking for some years before that, I still consider myself an amateur. Relatively few ornamental lathes are known to exist, and most of them are inoperative because of broken or missing parts; others are in museums or in the possession of private antique collectors and are not in use. Therefore, it is not strange that I, even with my experience, consider myself an amateur, as there are no schools or teachers and few master craftsmen to consult for instruction.

The nineteenth-century Holtzapffel family, who made more ornamental lathes than anyone else, wrote a five-volume set of books called *Turning and Mechanical Manipulation*; volume five was a complete description of the ornamental lathe and its operation. These books, however, are difficult to read and understand by an amateur turner even if he or she has been able to acquire a lathe; they seem to have been written by engineers for other engineers. Many amateur ornamental turners, including myself, are not engineers, and the Holtzapffel books seem unduly difficult.

Having struggled through the books, learned by the do-it-yourself method, and acquired a modicum of proficiency, I thought compiling an easy-to-follow book would be a good idea, helpful to those who appreciate the craft, either as practitioners or as collectors.

It is hoped that this book will create interest in ornamental turning and perhaps increase the proficiency of others who practice this fascinating craft.

—Frank M. Knox

ORNAMENTAL TURNERY

ONE

Ornamental Turnery: An Introduction

FROM the time of the Renaissance, a series of craftsmen developed a proficiency for working in ivory on a machine known as a lathe. The lathe spun the ivory around in place as the craftsman applied a chisel to its surface to create shapes, sometimes of extreme beauty and sometimes of queer and bizarre forms. Then, removing the basic form from the lathe, the craftsman applied hand-held chisels and files to decorate and embellish the surface of the piece further.

This was not the earliest use of a lathe to help form ornamented objects. It is reported that Phidias, before the fifth century B.C., at times used the lathe for the primary shaping of materials and then, with consummate skill, added additional decoration and embellishment to create some of the artistic pieces for which he became famous. From the earliest use of the lathe, however, until the nineteenth century, the basic means for embellishing a turned piece did not change: surface embellishment could not be applied while the ivory was on the revolving lathe; the pieces had to be removed and decorated by hand. Not until after the Holtzapffels perfected the ornamental turning lathe in the nineteenth century could the ornamentally decorated pieces be completed while still on the lathe.

Ornamental turnery is the decoration or embellishment of plain-turned objects, usually of wood or ivory, with designs that elevate them from merely utilitarian objects to bring them into the realm of the decorative arts. Decorative arts have traditionally been created in such media as glass and ceramics, which lend themselves to surface embellishment during the formation of an object. Although many materials may be used for ornamental turning, including metal, some types of rock, and plastic, the two primary materials used are wood and ivory. Neither of these is as malleable as such materials as blown glass or clay and therefore do not easily allow surface decoration during the forming or turning process. Therefore, some technique for applying surface decoration of ivory and wood during the turning process had to be developed.

This problem was overcome with the development of the ornamental-turning lathe, with its supplement of complex subsidiary tools and apparatuses. A plate, chalice, vase, candlestick, or other object may be turned on an ordinary lathe and become a thing of beauty by virtue of its material, form, and craftsmanship, despite its plain shape. The same object, turned on an ornamental lathe, may become even more beautiful by virtue of the surface decoration imparted by the equipment and tools.

Ornamental turnery as practiced by the medievel turners largely ceased to exist with the passing of the medieval period, but it was not completely forgotten. Although examples of the craft can be seen in many of the world's museums, it is almost impossible to find a specimen in even the finest art and antique shops for purchase by contemporary collectors. Practically all such existing work is either in museums or in the hands of private collectors. The work may still be seen in such places as the Pitti Palace of the Museo degli' Argenti in Florence, the Bayerisches Museum of Munich, the Musée du Conservatoire des Artes et Métiers in Paris, the Metropolitan Museum of Art in New York, and the Rosenbourg Castle in Copenhagen. For the most part, these are ivory works created by European craftsmen who lived in southern Germany, particularly Nuremberg, from the seventeenth through the nineteenth centuries.

Figures 1 through 6, examples of work in ivory from leading museums of the world, illustrate

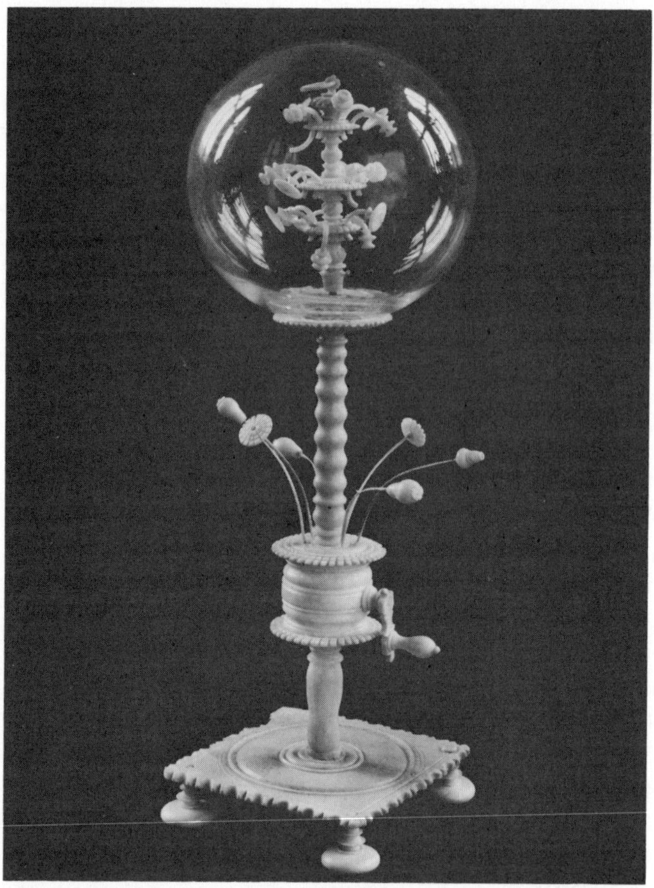

Fig. 1. Ornamentally turned piece by an unknown turner, seventeenth or eighteenth century. When the handle is turned, the flowers in the sphere revolve. (*Collection of the Bayerisches Nationalmuseum, Munich*)

Fig. 2. Exquisite interlocked piece by Jean Barreau of France, 1799. (*Collection of the Musée du Conservatoire National des Arts et Métiers, Paris*)

Fig. 3. Stemmed chalice in ivory, attributed to the Zick family. (*Metropolitan Museum of Art, Gift of Robert Gordon, 1910*)

Fig. 4. An ivory vase, 1624, by Johann Eisenberg, who worked under the patronage of the Medicis. A companion piece bears the inscription, "Art goes now from house to house asking bread. Oh merciful God; the bread will then aspire to art, but I will not come to profit by it." (*Pitti Palace, Museo degli Argenti, Florence*)

Fig. 5. Four pieces of ornamental turnery done in ivory by an unknown craftsman, probably in the seventeenth or eighteenth century. (*Collection of the Bayerisches Nationalmuseum, Munich*)

the incredible techniques that were developed during the sixteenth and seventeenth centuries. These pieces were not truly ornamental turnings because they were done partly on the lathe commonly used at the time and partly by hand decoration after the piece had left the lathe. *Mechanick Exercises or the Doctrine of Handy-Works* by Moxon and *L'Art du Tourner* by Plumier, as well as Diderot's encyclopedia, show numerous pictures of the apparatuses they used, none of which are as complete as the later Holtzapffel lathe. Moxon, in particular, not only shows pictures of these lathes, but describes them in great detail, discussing the pole lathe, the great wheel lathe, the treddle lathe, and the string lathe, as well as the use of these lathes in the turning of exotic pieces such as globes or balls, balls within one another, balls with a solid ball in the center, globes with several loose spheres in them, cubes within a hollow globe, oval shapes, and other such pieces.

One such special lathe used during this time was known as the swash plate lathe (fig. 7). A predecessor of the rose engine lathe, it had an attachment that moved the work back and forth along its axis with each turn of the lathe. This movement produced work somewhat like that

Fig. 6. Ivory box with cover and chain. The piece on the right bears the signature "Princeps F MDLXXVIII" inside the lid and is the work of Prince Ferdinando under the guidance of Sengher. (*Pitti Palace, Silver Museum, Florence*)

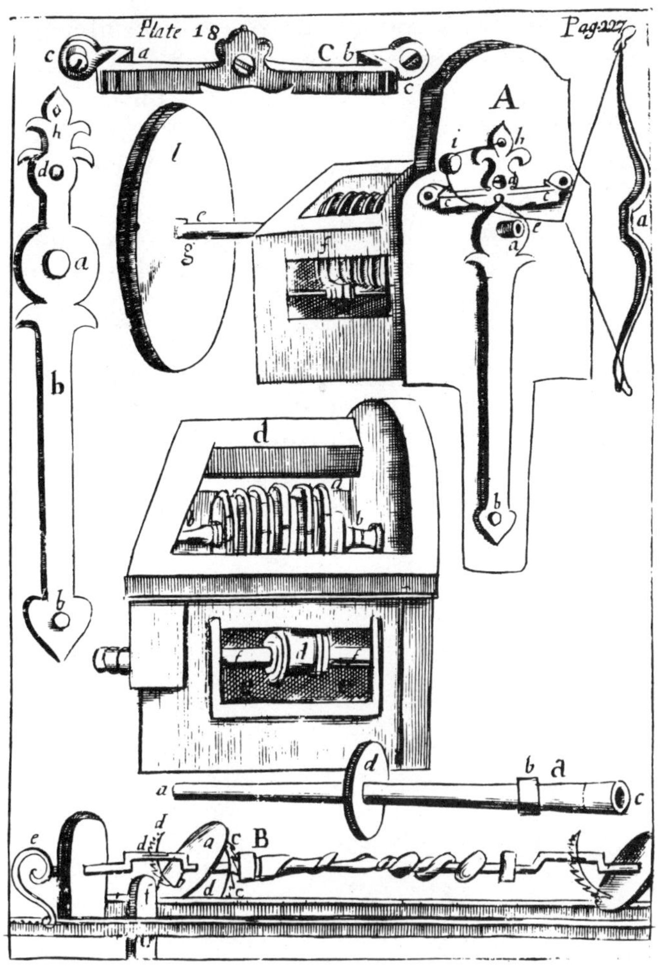

Fig. 7. One of the earliest known types of ornamental-turning lathes, the swash plate lathe. The swash plate is shown at *a* on the shaft *B*. It moves the work laterally back and forth along the lathe bed. At *d* is a primative type of rose engine lathe. (*Joseph Moxon,* Mechanick Exercises or the Doctrine of Handy-Works, *1703; Reprinted by the Early American Industries Association*)

from the rose engine lathe (see figure 14), except that the variations in cuts were tilted right and left horizontally along the axis of the work rather than toward and away from the central point of the work as it revolved in a perpendicular plane. Holtzapffel and other nineteenth-century lathe makers produced rose engine lathes but, so far as we know, did not make swash plate lathes.

The apparatus for ornamental turnery underwent much refinement in the nineteenth century, as will be discussed in more detail in later chapters. Although the craft declined with the coming of industrialization in the late 1800s, it is still practiced today by devoted enthusiasts. Anyone who has done only plain turning will be amazed by the creativity possible in ornamental turning. The ornamental-turning lathe is one important element in this increased creativity, but just as important is the turner himself. Without some sense of design, without a devotion to perfection of workmanship, without an affinity to creative development, all the equipment in the world will not produce acceptable results.

The development of this form of the creative arts opened the door to almost endless possibilities for embellishment and decoration of the surfaces of the object, sometimes with results that are wonderfully beautiful, sometimes unbelievably bizarre. Some turners decorate every bit of the surface merely because the instrument makes it possible to do so. More often than not this clutters the surface so much that the viewer has no opportunity to put the design into perspective. The possibilities of decorative patterns, using the bewildering variety of chisels, chucks, drives, and other attachments that have been developed for the ornamental turner, make it only too easy to overdo the embellishment of the surface. It is the responsibility of the craftsman to prevent overdesign. If the craftsman has any real sense of design and art appreciation, he or she can devote years to the lathe and never overdo a design.

Today there are no professional ornamental turners as the craft is too little known. Few if any ornamental turners have grown up with the craft as was the case of the craftsmen of the Renaissance period who learned ornamental turnery as youthful apprentices. Because ornamental turners are not professionals, there are no established standards by which to measure skill and performance. The amateur of today can measure his performance only by generally accepted standards of the creative arts and the performance of his peers.

Just as one original painting, however modest, carries something of the artist's spirit, so any handmade object brings with it something of the delight of its maker. For the ornamental turner, there is no enemy other than time; life is all too short for learning the job and doing it well. He cannot start too early nor live too long.

TWO

The Beginnings of Ornamental Turnery

No one knows for certain when plain turnery was first practiced, and the origins of ornamental turnery are also not positively known. The earliest known example of plain turning is thought to be a fragment of an Etruscan bowl from a tomb in Corneto, from about 700 B.C. As mentioned earlier, it is thought that Phidias used a lathe for some of the work on his sculptures in about 500 B.C. The earliest known illustration of a lathe is a carving on the wall of an Egyptian tomb from about 300 B.C. Pliny ascribed the invention of the lathe to Theodore of Samos, whereas Diodorus Siculus (50 B.C.) earlier reported that the first person to make a lathe was a nephew of Daedalus of Greek mythology. Turning was probably introduced into England by the Iron Age Celtics around 200 B.C. Regardless of the actual date, there is no question that the turning lathe was the earliest known mechanical aid to the woodworker.

Because of the paucity of written records, little is known about the use of the lathe during medieval times. Not until the Renaissance does evidence of turning again appear. Chartres Cathedral, built in the thirteenth century, has two stained-glass panels that show turners. Leonardo da Vinci (1452–1519) designed three lathes, one with crank and handle; Gio Paulo Lomazzo described Leonardo's oval turning in 1590. In 1703, in his book *Mechanick Exercises or the Doctrine of Handy-Works*, Joseph Moxon described turnery as follows:

As by placing one Foot of a pair of Compaſſes on a Plane, and moving about the other Foot or point, deſcribes on that Plane a Circle with the moving point; ſo any Subſtance, be it *Wood*, *Ivory*, *Braſs*, &c. pitcht ſteddy upon two points (as on an *Axis*) and moved about on that *Axis*, alſo deſcribes a Circle Concentrick to the *Axis*: And an Edge-Tool ſet ſteddy to that part of the outſide of the aforeſaid Subſtance that is neareſt the *Axis*, will in a Circumvolution of that Subſtance, cut off all the parts of Subſtance that lies farther off the *Axis*, and make the outſide of that Subſtance alſo Concentrick to the *Axis*. This is a brief Collection, and indeed the whole Sum of *Turning*.

The first encyclopedia, prepared by Denis Diderot D'Alembert from 1751 to 1772, illustrates lathes and the work done on them. Other historical references describe lathes and turning as well; this summary indicates the antiquity of the craft and its place in human history.

Royal Patrons and Promoters of Ornamental Turnery

The practice of the turner's art was very popular during the Renaissance. The rulers of the Holy Roman Empire figured largely as sponsors, not only of the craft in general but of individual craftsmen as well, some of whom later became outstanding leaders in the field.

From the sixteenth through the eighteenth centuries there was a surge of interest in ornamental turnery due largely to the increased production and availability of lathes during this period. No exact count of lathes is available but it must have been in the scores if not in the hundreds. Most of them were of the rose engine variety upon which were made some of the bizarre shapes shown in figures 1 through 5. Other lathes include the swash plate, the copying lathe, the foot treadle, the bow lathe, as well as just plain, simple lathes. The eighteenth century was the most prolific in the production and use of ornamental-turning lathes and furnished the basis for the much-improved lathes made in the nineteenth century by Holtzapffel.

Among the patrons of the Holy Roman Empire were: Maurice, elector of Saxony, who reigned from 1547 to 1553; Emperor Rudolf II of Bohemia, 1552 to 1612; George William, elector of Brandenburg, 1619 to 1640; Ferdinand III, king of Hungary from 1625 to 1657 and Holy Roman emperor from 1637 to 1657; Frederick Augustus, elector of Saxony, 1694 to 1723; Joseph II, Holy Roman emperor from 1764 to 1790; as well as John William of the Palatinate and Ferdinand of Bohemia.

In the sixteenth and seventeenth centuries, a dynasty of ornamental turners lived in Nuremberg. The Zick family consisted of the father, Peter (1571–1629), his son Lorenze (1594–1666), and his grandson Stephen (1639–1715). Stephen, master turner to the emperor, was responsible for the design and execution of many of the exotic and seemingly impossible pieces seen in museums today.

Rulers themselves often practiced turnery. The Brandenberg in Copenhagen has an ivory case with a pair of compasses turned by Peter the Great. Also at the Brandenberg are boxes made by Kings Frederick III and IV of Prussia.

Peter the Great, czar of Russia from 1682 to 1725, was also a patron of turnery. Andrei Konstantinovich Nartov was an active state counselor and mechanic to Peter, as well as instructor in the art of turnery and a member of the Imperial Academy of Sciences. Nartov lived for twelve years at Peter's court and became the world's first machine-tool designer. He invented and developed slide rests that greatly expanded the ability of the lathe to create the more complicated forms that later became known as ornamental turnery. In 1718 Peter sent Nartov to Berlin, where he spent six weeks instructing Frederick William I, king of Prussia, in the art of ornamental turnery. In 1719 Nartov went to Paris, where he found workmen capable of making two lathes from his drawings at a cost of 6000 livres. Another of Nartov's lathes, made in Moscow and brought to Paris, was presented to Louis XV with a collection of turned objects that were placed in the Louvre and are still kept, together with the lathe, in the Musée du Conservatoire National des Arts et Métiers.

Professor A. S. Britkin, in his treatise "Nartov and His 'Theatrum Machinarum,'" wrote:

> All the works produced by Nartov are of outstandingly complicated shapes and delicate workmanship; goblets with lid, oval vessels and other curious shapes, constituting a combination of convex surfaces, convolute areas and complicated figures all placed in fanciful configurations. It is possible to shape the material into any form and combination of forms limited only by the imagination. With delicate filigree tracery there were polygonal spiked bodies with intricate detail, series of rings made on the lathe and joined with small triangular pyramids, through rings protruded conelike needle-shaped elements mounted inside the structure, hollow ivory bodies of various shapes, etc.

In the mid-nineteenth century, Nicholas I, czar of Russia, presented the Austrian emperor Ferdinand with a lathe made by Nartov, which is now located in the Technical Museum of Vienna.

Christian V, king of Denmark from 1641 to 1690, and Frederick IV, king of Denmark from

1771 to 1830, made objects that are now in the Copenhagen Museum. Queen Juliane Marie of Denmark made a complete ivory temple in miniature. Kings Louis XV and XVI of France were both turners and collectors. Princess Louise, daughter of George II of England, made a complicated pyramid. At Augsburg are the pieces of Tabien Treffler, who taught ornamental turnery to the two daughters of the duke of Brunswick. Ornamental turnery was introduced into Denmark by the Swiss master Lorenz Spengler.

Post-Renaissance Turnery

Technically and artistically the French led the field of post-Renaissance turners. Charles Plumier and Louis-Georges-Isaac Salivet, authors of early books on the subject in 1689 and 1792 respectively, were outstanding craftsmen. An important center of turnery developed around Meru in the Oise. The Musée du Conservatoire National des Arts et Métiers in Paris has examples of the work of the great French turner Barreau, who was born in Toulouse in 1731 and died in Paris in 1814. As a young man, Barreau went to Avignon, where he learned his trade in the workshop of Jean Berard.

Ornamental turnery was also practiced in England, particularly in the nineteenth century, when John Jacob Holtzapffel resided in London. Holtzapffel's accomplishments will be described in more detail in chapter 3. The high cost of the lathes in post-Renaissance England limited their use in large part to royalty and the peerage. Some army officers bought lathes and took them when they went on assignments to the far reaches of the British empire. It is interesting to note that British women also practiced turning. Two of these turners, Lady Gertrude Crawford and Ellen Willmott, were very highly regarded; indeed, Lady Crawford is honored even today by the annual awarding, by the Worshipful Company of Turners of the City of London, of the Lady Gertrude Crawford Silver Medal for excellence in ornamental turning.

At the famous Wedgwood Pottery Works in Staffordshire, England, is a rose engine lathe that was apparently made prior to April 1764. As recently as 1954, it was in continuous use, although it has not been displayed since then.

Collections Around the World

The collection of ornamental turnery sponsored by the Medicis and now housed in the Pitti Palace in Florence (see figs. 5 and 6) is a fantastic array of shapes and materials. The Victoria and Albert Museum has a piece turned by Cosimo de' Medici III himself.

The Science Museum of London has a square Gothic church tower, 21 inches high, crafted by John Jacob Holtzapffel II in ivory. It appears to be a replica of Chartres Cathedral or Westminster Abbey, complete with arched windows, delicate tracery, and all the ornate trimmings of Gothic architecture. It was made completely on the ornamental lathe in separate pieces and then assembled.

A drinking cup made by Prince Elector John George of Saxony, who ruled from 1585 to 1657, has survived and is on exhibit in Vienna. In a room between the Winter Palace and the Hermitage Museum in Leningrad are several lathes together with specimens of work in ivory brought from Holland by Peter the Great.

Ornamental turnery has little history in the United States. In Buffalo, New York, the Museum of Natural Sciences has a collection of objects turned by a Michael Solomon in the late nineteenth century, but there is no record of the lathe on which he worked and little is known about him. His work included hollow ivory balls that contain six- to sixteen-pointed stars and other incredible forms, locked within the balls and inseparable from them.

The world's museums contain many pieces of ornamental turnery done by masters of the craft, some sponsored by royalty and some done by members of royalty themselves. But these were done in past centuries; little work of museum quality was done in the first half of the twentieth century. It seems that the industrial revolution buried this extraordinary craft until a small group of men in England resurrected it in the mid-twentieth century.

Books and Treatises on Ornamental Turnery

Little has been written on ornamental turnery. Some of the most important treatises were prepared between the seventeenth and the nineteenth centuries and can be found only in bookstores that sell old, rare books.

In the late seventeenth century, a friar of a mendicant order, Charles Plumier (1646–1704), wrote the first manuscript on ornamental turnery. He completed it in 1689 and sent it to the head of the order in Rome, where it was approved by the pope. The first edition of Plumier's work was issued in Lyon in 1701 and was reissued in Paris in 1706. The second edition was issued in Paris in 1749. Peter the Great of Russia ordered a Russian and Dutch translation in 1716. This manuscript is listed under Peter the Great in the manuscripts catalog of the Academy of Sciences in Leningrad. It was never published.

In 1974 Plumier's manuscript was translated into English by Dr. Paul Lawrence Ferraglio of Brooklyn, New York, and privately published in a very limited edition. Both the original Plumier, known as the "L'Art du Tourner," and the English translation are now rare and eagerly sought by bibliophiles and turners.

In 1703 the Englishman Joseph Moxon wrote *Mechanick Exercises or the Doctrine of Handy-Works: Applied to the Arts of Smithing, Joinery, Carpentry, Turning, and Bricklayery*. Moxon was a Fellow of the Royal Society and an outstanding leader in the dissemination of knowledge of the crafts. His book has, fortunately, been reprinted by the Early American Industries Association and is now available to modern craftsmen. Moxon's book and that of Plumier were standard works on turnery until superseded by that other French classic, the *Manual du Tourner*.

Andrei Konstantinovich Nartov, of Peter the Great's court, wrote the treatise "Theatrum Machinarum," but it was never published. It still exists in manuscript form and has been translated from the Russian by the Israel Program for Scientific Translations in Jerusalem in 1966. Edwin A. Battison of the Smithsonian Institution inspired this translation; it was sponsored by the Smithsonian and without that support probably never would have appeared.

Another book on ornamental turnery, written by Louis-Georges-Isaac Salivet under the pseud-

Mechanick Exercises.

OR THE

DOCTRINE

OF

HANDY-WORKS.

Applied to the Arts of { Smithing Joinery Carpentry Turning Bricklayery.

To which is added

Mechanick Dyalling: Shewing how to draw a true Sun-Dyal on any given Plane, however Scituated; only with the help of a ſtraight *Ruler* and a pair of *Compaſſes*, and without any *Arithmetical Calculation.*

The Third Edition.

By JOSEPH MOXON, *Fellow of the Royal Society, and Hydrographer to the late King* Charles.

LONDON:

Printed for *Dan. Midwinter* and *Tho. Leigh*, at the *Roſe and Crown* in St. *Paul's-Church-Yard.* 1703.

onym Louis-Eloy Bergeron, is commonly known as the Bergeron work. It was written in 1792 as a two-volume set and was supplemented in 1816 by a third volume written by Pierre Hammelin Bergeron, Salivet's son-in-law. This third volume consists of magnificent steel engravings of the lathe, its attachments, and the work produced by turners of the time.

This volume introduced me to ornamental turnery. I discovered a copy in a Los Angeles bookshop in a very bad state of disrepair, with the leaves and plates intact but the binding almost completely gone. The book dealer did not know what it was, since neither he nor I had ever heard of ornamental turnery at that time. Because I collect books and fine prints, I bought the book and had it bound. I pursued the subject by sending a query to the International Wood Collector's Bulletin and received an instant reply from a member of the Society of Ornamental Turners in England, which set in motion a chain of events that ultimately resulted in my obtaining a lathe in 1963, which I still use today.

John Holf Ibbetson's *Geometric Chuck*, published in 1833, is a treatise on the Holtzapffel geometric chuck and its uses. He makes a very interesting comment on its use as follows: "The object of this memoir is to bring to the attention of the Public—particularly the amateur turner and those who take pleasure in the investigation of the organical description of curves—the powers and capabilities of the instrument alluded to in the title page." This gives the beginner some idea of the esoteric nature of some of the lathe attachments.

The most significant ornamental turning publication to date, the five-volume set of books *Turning and Mechanical Manipulation* by the Holtzapffel family, is a monumental work on the subject of ornamental turnery and its related fields published between 1843 and 1884. Volume five of the set is a definitive treatise on the ornamental lathe, with descriptions and detailed illustrations of every aspect of the lathe and its use. Volume four contains introductory material on simple turning, much of which can be directly applied to ornamental turnery as well. The first three volumes deal with related subjects, but to the ornamental turner the last two volumes are bibles. These, fortunately, have been reprinted by Dover Publications of New York. The original volumes command quite a high price in the rare book field.

Unfortunately for the amateur turner, the Holtzapffel volumes seem to have been written by engineers for other engineers and not for ordinary amateur turners. The very stylized writing, combined with extremely long and involved sentences, both common to the period, are difficult to follow. Because so little has been written about ornamental turnery, this set still provides the ornamental turner with the most complete guidance. The lack of experienced ornamental turners in the United States means that all too often beginners learn the craft on their own, by trial and error, going back again and again to Holtzapffel to reread in order to find what was missed in earlier readings.

In 1857 Captain James Ash, inventor of the elliptical cutting frame, wrote *The Art of Double-Counting on the Lathe*, published by L. Booth, 307 Regent Street, London. This was a highly technical treatise for those using his invention.

FIG. 8. The five-volume set of books written by the Holtzapffels. The individual volumes were published under several different titles but they constitute a homogeneous set and are commonly known and accepted under the name *Turning and Mechanical Manipulation.* Volume five is the bible of the ornamental turner. This original set was purchased in 1965 with the covers practically gone. It was rebound and slip-cased. The interior sheets were fortunately in excellent condition.

The Reverend James Lukin (1827–1917), one of the outstanding ornamental turners of the modern period and an accomplished engineer, wrote several books, principal of which was *The Lathe and Its Uses*. I consider this book second only to Holtzapffel in usefulness. The first edition was published in 1868, and the fifth edition was published in 1878. Lukin clarifies much of what Holtzapffel leaves unclear. Lukin also wrote three other volumes: *A Manual for Technical Schools and Apprentices*, fourth edition, 1894; *Possibilities of Small Lathes*, published in 1903; and *Simple Decorative Lathe Work*, published in 1905.

A few other books on ornamental turnery exist, but these representative titles indicate the scarcity of information on the subject. Undoubtedly the most prolific source of information about the ornamental lathe and its use is to be found in the bulletins of the Society of Ornamental Turners. At the time of this writing, seventy-four bulletins had been issued, on a semiannual basis. Their distribution is limited to members of the society, and they have never been reprinted or offered publicly. In their pages is to be found a veritable wealth of information and opinions on practically every facet of ornamental turnery.

THREE

Ornamental Turnery Comes of Age

IN the nineteenth century, ornamental turnery matured with the innovations perfected by the Holtzapffel family and others. With the production of more than 2,000 greatly improved lathes, this was the golden age of ornamental turnery.

The Holtzapffel Firm

The beginnings of true ornamental turnery are found in the story of John Jacob Holtzapffel, an Alsatian engineer who settled in London in 1792, while George III was king of England. From the first his main objective seems to have been producing lathes and related tools. In 1795, he delivered the first Holtzapffel lathe to a Mr. Crisp. By the end of 1803, he had made and sold no fewer than 358 lathes.

Prior to Holtzapffel's work, ornamental turnery really did not exist in the technical sense, as most of the ornamentation had to be applied by hand after the piece was plain-turned. Holtzapffel developed the overhead drive, which operates much like the modern dentist's drill. He also designed a myriad of revolving cutters that operate in a slide rest to produce the ornamentation that previously had been done by hand.

In 1804 the firm was expanded to include John George Deyerlein, but the arrangement was not happy, and Deyerlein left the firm in 1827. In the same year Charles Holtzapffel, John Jacob's son, joined the firm. Charles was a distinguished engineer in his own right, even though he was only twenty-one when he joined the firm. John Jacob Holtzapffel died in 1835. The firm was carried on by Charles, who published his father's first volume of *Turning and Mechanical Manipulation* in 1843. The second volume was brought out in 1846.

Charles Holtzapffel established himself as a maker of mechanical apparatuses for amateurs and continued his father's work in developing the machinery and various attachments for ornamental turnery, as well inventing other devices. Charles developed machinery for printing banknotes, a tool for cutting rosettes for ornamental turning, a dividing engine for the graduation of drawing scales, and an apparatus for tracing geometrical figures on glass. It was said that he probably never put his hand to a machine that he did not improve in some manner or other. Charles died in 1847 and his widow, Amelia Vaux (Dutton) Holtzapffel, managed the firm until January 1853.

Volume three of *Turning and Mechanical Manipulation* was published by Amelia from Charles's manuscript in 1850. In 1867 Charles's son, John Jacob Holtzapffel II, became head of the firm and continued his father's efforts with the publication of volume four of the set in 1879, following with volume five in 1884. The set was not complete, however, until a revised and enlarged edition of volume three followed in 1894.

Charles Holtzapffel's nephew, George William Budd, later known as George William Holtzapffel, became head of the firm in 1896. John Jacob Holtzapffel II died in 1897.

The Holtzapffel firm still existed in the early years of the twentieth century, Colonel John George Holtzapffel Budd, son of George William, having joined in 1919. But very few lathes of the sort now beloved by ornamental turners were made after the nineteenth century, and that century is rightly known as the high point of the ornamental-turning lathe.

During the Napoleonic Wars (1793–1815), the Holtzapffel firm delivered 110 lathes to the Royal Arsenal. During the American Civil War, the Confederacy bought a rose engine lathe for engraving its postage stamps, but since it never paid the bill, it is assumed that the ship carrying the lathe never ran the Union blockade.

It is interesting to note some of the prices at which the Holtzapffel lathes were sold at the time of manufacture. The record is very incomplete, such records as are available having been taken from the Register of Lathes 1795 to 1918, MS 9475, in the Guildhall Library, London.

The earl of Harborough was undoubtedly the best customer for Holtzapffel lathes, having bought no fewer than nine during his lifetime. The first was #855, bought in 1812, and the last was #1934 in 1848. Only two of these nine lathes are known to exist today.

Lathe #1972 was ordered in July 1849 for the 1851 Great Exhibition of the Works of Industry of All Nations. It was a superior lathe in all respects, with much auxiliary apparatus, mounted on a base of beefwood. It was on exhibition until October 11, 1851, when it was sold to Richard, fifth baron of Berwick, who died in 1861. The next owner of record was Major Robert Chadwick of Findhorn, near Forres, Morayshire, who paid £200 for it. It was kept in the Chadwick family for several generations until being presented to the Colchester Museum in Essex.

In 1871 lathe #2287 was sold to Captain W. H. Roberts for £535; in 1891 it was resold to Ellen Ann Willmott for £425:

> Mrs Willmott had misgivings about bits and pieces of machinery on the drawing-room tables ruining their surfaces, or perhaps Ellen herself preferred to work in comfort in a room where she could do as she liked, keep everything where she pleased, and above all concentrate without visitors peering over her shoulder as she worked: for it was indeed work which demanded a sure, deft hand and the worker's full attention. So the work-room came into being, and it was always kept locked. Mrs Willmott perhaps reconciled herself to the presence of this machine, and the frequency with which her daughter shut herself up in the work-room, by reading the list of distinguished female turners: as these included Lady Emily Fitzmaurice, Margaret Lady Amherst, the Marchioness of Ormonde, and Baroness Burdett-Coutts, to say nothing of many titled owners going back to Marchioness Townshend, who was the first of the long

Fig. 9. Ornamental turning lathe by Holtzapffel, #2023, made in 1853. I motorized it in my workshop, where I have used it since 1963. Not shown are the many attachments and auxiliary parts.

Fig. 10. The slide rest for Holtzapffel lathe #2023. The primary attachment to the lathe, it is removable and can be adjusted to accept other attachments. Shown is the belt from the overhead drive activating a horizontally revolving cutter. (*Photo by John Kelsey for* Fine Woodworking)

Fig. 11. The author at work on his Holtzapffel lathe. In the tool holder is the eccentric cutter making the barleycorn pattern. (Photo by George Erml)

line in 1798, then perhaps Ellen could have chosen a worse occupation. Mrs. Willmott must also have closed her eyes to the cost of further apparatus ordered by Ellen in December 1892 (or perhaps she had simply given up trying to curb her daughter's extravagance). The description of the Rose Chuck, Rose Engine and Rose Cutting Frames and Patent Automatic Driving Gear and Segment Stop Apparatus (£90) plus packing case and packing (7s 6d) occupied six full pages of Messrs Holtzapffel's ledger, and kept Ellen happy until 1897, when she again launched forth, this time into the purchase of a new pattern balanced Eccentric cutting frame and a further Rosette for use with the Rose Chuck, at a cost of £21 5s.

With this apparatus she produced some extremely attractive turnery, and is said to have made rings out of coins and performed other party tricks, besides the more serious and accomplished work which she produced in ivory and rare hardwoods. The delicate ivory boxes she made were much sought after as wedding presents. She seems to have been extremely neat-fingered and to have possessed an unusually well-developed aptitude for handling machinery. Perhaps her naturally inquiring mind directed her fingers.

In 1853 the Holtzapffel company completed work on lathe #2023 for some now-unknown buyer. For more than 110 years, the lathe was used, probably by a succession of owners, until in 1963 it came into my possession. This lathe is extraordinarily well preserved, of dark mahogany, machined steel, and polished brass. It is one of the few such lathes presently being used to produce ornamentally turned objects of art in rare and exotic woods.

Other Lathe Makers

Holtzapffel was by no means the only shop making ornamental lathes. Although it made far more than any other, several other firms made lathes and some very good ones at that.

EVANS LATHE. The best-known lathe maker contemporary with Holtzapffel was Evans. The shop made fewer lathes, but they were nevertheless fine specimens, with modifications not found on the Holtzapffel lathes. William Jones Evans was an engine, lathe, and tool maker in London. He established his firm in 1810 but finished his working life with Britannia Company of Colchester.

BIRCH LATHE. Only a few ornamental lathes were made by the George Birch & Company of Salford, Manchester, England. Most of these lathes were engineers' metalworking tools for precision work. They were given full ornamental turning capability with the addition of the necessary components but, unlike those of Holtzapffel and Evans, were never made originally as ornamental lathes. One Birch lathe was reportedly made in 1888 for a Reverend C. C. Ellison for a total cost of £860, the equivalent of about $3,000 today.

PLANT LATHE. The firm of George Plant made ornamental lathes, one of which was uncovered in an old shed by a member of the Society of Ornamental Turners and purchased. After it was cleaned up, the inscription "George Plant, maker, Kidsgrove, Staffs, 1954" was uncovered.

HULOT LATHE. A French toolmaker, Hulot, made a rose engine and medallion-copying lathe, probably in the middle part of the eighteenth century, which found its way to England. It seems that George III summoned Hulot to England in 1766 in connection with an engine-turning and medallion-copying lathe which the king had on order. James Watt went to Paris and saw such a lathe in the workshops of Hulot.

GOYEN LATHE. William Goyen made a small number of high-quality lathes for amateur use in Newton Abbot, England, in the last quarter of the nineteenth century. Little is known about Goyen, a South American railway engineer who retired and took to making lathes as a hobby. He is credited with making singlehandedly the finest ornamental lathes. They were probably made for friends; two of them were made for the Singer brothers of sewing machine fame for a reputed cost of £1,500 each. Only ten Goyen lathes are known to exist today.

MUNRO LATHE. James Munro was an engineer, machinist, and lathe and tool maker who worked

FIG. 12. An ornamental lathe made in the nineteenth century by James Munro, a former employee of the Holtzapffel Company. It has a traversing mandrel for screw cutting. (*Rev. James Lukin,* The Lathe and Its Uses, *published by Trubner & Co., London, 1878*)

in London. He was a maker of ornamental lathes, as is evident from figure 12, a lathe dated 1868. The slide rest on my Holtzapffel lathe is engraved *J. Munro, London.* An 1860 advertisement headed "James Munro (From Messers, Holtzapffel & Co.)" would indicate that Munro had been an employee of Holtzapffel before setting up in his own business. The number of ornamental lathes that Munro made is not known. The quality of his lathes, as illustrated by the screw-cutting lathe shown in figure 12, would seem to be equal that of known Holtzapffel lathes.

LUKIN LATHE. Catalog number six of the Britannia Company, Colchester, England, shows a picture of the Lukin ornamental-turning lathe. It is similar in construction to the other nineteenth-century lathes. As mentioned in chapter 2, Reverend James Lukin was an accomplished turner and engineer who wrote several books on turning.

FENN LATHE. Joseph Fenn was another Holtzapffel employee who left the firm to start his own toolmaking company. The Reverend G. A. Grace, a member of the Society of Ornamental Turners, reported, "I have a very nice Fenn lathe, the workmanship of which is in every way of the Holtzapffel standard."

Undoubtedly other ornamental lathes were made by other toolmakers, but the above are the better-known ones. No company ever matched the productivity of the Holtzapffel family in the field of ornamental lathes.

The Society of Ornamental Turners

A memorable event in the history of ornamental turnery took place in London on April 20, 1948. Six gentlemen, S. G. Abell, W. G. Hindes, H. E. Heald, S. G. Askey, F. J. Howe, and A. Purdie, gathered at the Science Museum in South Kensington to view a display of Holtzapffel lathes.

These gentlemen all were interested in ornamental turnery and the lathes made by Holtzapffel as well as other ornamental-turning lathes. The collection of Holtzapffel machines in the Science Museum was the prime objective of their visit, but a note in the display case indicated that a reserve collection might be of further interest. After some discussion with the curator an invitation was issued to see the reserve collection at a future date. After seeing the Holtzapffel lathes, the party adjourned to a nearby café for tea and a chat, the result of which was the formation of an unofficial committee, whose subsequent work led to the formation and inaugural meeting of the Society of Ornamental Turners. Norman Tweddle, who already had forty years' experience as a ornamental turner, joined the informal committee and became the first president of the organization. Its first meeting was held on October 23, 1948. The Society's first bulletin was issued on August 1, 1949, and listed as officers Norman Tweddle as president, S. G. Askey as vice president, S. G. Abell as treasurer, and F. J. Howe as secretary.

Norman Tweddle, who died at the age of sixty-four in 1953, was one of the great machinists and craftsmen of his time. Mechanics, especially ornamental turnery, were the abiding passion of his life on his retirement, and his Holtzapffel lathe, #2185, which he restored with meticulous care, became one of the finest extant. He also owned a Holtzapffel rose engine lathe, #16, and went to work on the geometric chuck, developing it beyond the original intent of the makers. He obtained a Hartley chuck and Hartley's treatise on it, which was published by the society. He is generally conceded to be one of the best, if not the best, of the modern ornamental turners. He successfully worked out the mechanics of the famous "Chinese Balls," designed and constructed the necessary tools to make them, and completed several sets of this difficult piece entirely by hand. The "Chinese Balls," or celestial spheres, are made of ivory and consist of spheres set inside of each other, with each sphere moving independently of its neighbors. The number of spheres ranges from three or four to as many as twenty or more. In some cases the surfaces of the interior spheres are decorated by hand with carvings. They are extremely difficult to make and are some of the most sought-after pieces

Fig. 13. The founders and members of the Society of Ornamental Turners, London, 1948. They are: (*left to right*) Dr. S.G. Askey; A.V. Reed*; George Corderoy*; Brigadier Wood; Sydney Abell; W.G. Hindes; Harold Yeatman; Ken Fowler; Fred Howe; Frank Haythornthwaite; A.W. Jones*. (*Surviving member.)

of oriental art. The author has two sets in his collection.

Another notable turner of this period, also an active member of the society, was the Reverend G. A. Grace. In his eighties, the reverend spent several hours a day in his shop, using the foot treadle to power his lathe. He had a complete workshop including a Myford engineering lathe; both an Evans and a Fenn ornamental lathe; a Pittler all-purpose lathe; drilling, sharpening, and grinding machines; a workbench; and cabinets of tools. In a bulletin of the Society of Ornamental Turners, a visitor to the Grace shop reported:

> The Evans lathe is in operation cutting surface patterns of the rose engine type by means of the pumping action. For this the mandrel is turned for the finishing cut once in thirty five minutes via the flywheel and hand, operated slow motion with very much winding. By this means table mats in Perspex, brooches and other items in ivory, blackwood, etc., are being produced. Using a steel tool in a horizontal or universal cutting frame working at high speed the finish being obtained is near enough to perfection. Further experiments to combine this action with other apparatus are planned. As the reverend gentleman mildly remarks, "It helps to keep the mind lively."

On September 20, 1958, a tenth-anniversary meeting of the society was held at the Rotary House of Friendship. Thirty-one members and guests attended. Each visitor was handed a card that read:

> On behalf of the Council and members of the Society I have much pleasure in extending a hearty welcome to all visitors. The exhibition has been organized to mark ten years of achievement, nearly all of the exhibits having been made since the Society was founded in 1948. It is, I believe, the first exhibition entirely devoted to ornamental turning and certainly the most comprehensive ever held. The real merit of the exhibits lies in the fact that each is an individual production, a minor work of art.
>
> I hope very much that you will enjoy your first visit and think that the Society has justified its existence.
>
> Signed, S. G. Abell, President

The February 1959 issue of the *Bulletin of the Society of Ornamental Turners* carries a report of the tenth-anniversary meeting and a set of five photos of work exhibited by the members.

The bulletins of the Society of Ornamental Turners are a veritable gold mine of information on the subject. With the exception of the Holtzapffel volumes, such information will probably never again be made available to amateur ornamental turners. The highest possible praise is due to the founders and members of this society for the contributions they have made to this ornamental turnery.

The society was formed as, and has remained, a relatively small group with only a few hundred members. Its small size results from the relative obscurity of the craft after the end of the Victorian period and the esoteric nature of the work, as well as from the scarcity of the lathes on which the work can be done. The growing interest among craftsmen, as well as collectors, in ornamental turnery, however, indicates a rapidly expanding interest in the subject. The scarcity of lathes unfortunately remains a roadblock to the expansion of the craft.

Current Status of Ornamental Turnery

The status of ornamental turnery is best summed up by the almost invariable response whenever the subject is mentioned: "What's that?" Very few people in even the crafts field have heard of it. On several instances I have found pieces of antique ornamental turnery in the antique shops of New York's Second Avenue, but seldom has the shop owner been able to identify it as ornamental turnery. In two such cases, the owners have hazarded the guess that "it is something in plastic," the pieces being turnings in African blackwood.

Incomplete records indicate that fewer than 1,000 ornamental lathes are known to exist today. This figure varies as lathes are discovered in attics or piles of old or discarded machinery. Most such lathes are broken or, in many cases, have missing parts. Members of the Society of Ornamental Turners, most of whom are accomplished engineers or toolmakers, have spent months, even years, in restoring such broken lathes and making the missing parts. Rarely does a previously unknown lathe appear without having to be repaired, more often than not to a considerable extent.

Of the known operable lathes, relatively few are in actual use. A number are in such museums as the Smithsonian Institution, the British Science Museum, the J. Paul Getty Museum, the Musée du Conservatoire National des Arts et Métiers in Paris, the Technical Museum of Vienna, and others. But even more lathes are inoperative or not being used, being mothballed in the hands of collectors. Some collectors have two, even three, lathes but use none of them. This, of course, greatly reduces the number of lathes in actual operation and diminishes the output of those in operation. Only in recent years has the craft become sufficiently known to make currently produced objects collectors' items. Objects from the Renaissance and Victorian periods are avidly sought by knowledgeable collectors. Most contemporary turnings are produced by members of the Society of Ornamental Turners.

Tremendous opportunity seems to exist for an enterprising toolmaker to investigate ornamental turnery and make a few attachments for the modern wood-turning lathe that would enable plain turners to practice this exquisite craft. The availability of more lathes and the resulting increase in the number of skilled ornamental turners could restore the craft to its rightful place in the field of decorative arts.

FOUR

The Ornamental-Turning Lathe

THE ornamental-turning lathe begins as a simple wood-turning lathe. It differs in the wealth of attachments that make possible decorative embellishment not feasible with an ordinary lathe. The primary attachments are an indexing plate and a slide rest. The indexing plate, on the front of the pulley, contains rows of holes into which a fixed pin can be inserted to stabilize the lathe mandrel in a given position and keep the work from turning. This allows the turner to make cuts in the precise locations he or she chooses. The slide rest holds the cutting tool in position and provides a mechanical means of moving it from one position to another. In addition to these two attachments, numerous auxiliary apparatuses provide the means for making the ornamental cuts. All of these attachments are discussed in more detail later in this chapter.

Four basic types of motion are possible with an ornamental lathe:

1. The work turns while the cutting tool or chisel is held stationary by hand on the tool rest (plain turning).
2. The work is held stationary by the index pin while the cutting tool turns by means of an overhead drive.
3. Both the work and the cutting tool turn.
4. The work turns and, at the same time, moves laterally by means of the traversing mandrel; the cutting tool is still or in motion.

A fifth type of motion may be obtained with a rose engine lathe, which has a rocking headstock.

The mandrels of the ornamental lathe are of hardened steel and run in hardened steel bearings. There are no ball or roller bearings, as are found in modern plain-turning lathes. Norman Tweddle commented on this anachronism, saying, "Provided they are kept freely lubricated they seem to be everlasting. No provision is made for taking up wear. The thrust must be carefully adjusted so that the mandrel turns easily by hand without any end shake whatever. On being properly adjusted it should be possible to spin the stepped pulley for at least a dozen revolutions on being pulled round by hand."

Power for the ornamental lathe is provided by a foot treadle operated from beneath. The treadle activates a flywheel, which in turn activates the belts that extend from the flywheel to the stepped pulley in the headstock as well as to the overhead drive. The modern ornamental turner invariably motorizes his lathe, creating problems that are not found in the foot-treadled lathes. The higher and more sustained speeds at which the motorized lathe can be run create a problem in oiling the main bearings. Because the lathe was built to run at slow speeds, only the amount of oil needed to run at these speeds reaches the main bearings. At faster speeds, too little oil reaches the bearings, and they freeze. This problem can be solved with the installation of oil drip cups. I learned this the hard way when the main bearing in my headstock froze. Fortunately I had a friend who was an expert toolmaker, and he was able to unfreeze the bearing, repolish it, and install an oil drip cup that solved the problem.

Types of Ornamental Lathes

One of the earliest ornamental lathes was described by Joseph Moxon in his 1703 book *Mechanick Exercises or the Doctrine of Handy-Works*. This, the swash plate lathe, produced many eccentric shapes. The swash plate lathe, shown in figure 7, can be seen only in museums, but other lathes, described below, are all used by turners today.

SIMPLE ORNAMENTAL LATHE. This lathe has a nontraversing headstock mandrel that carries the chucks that hold the work just as an ordinary lathe does. With attachments, it can do all the ornamental cutting actions the average turner may need. It cannot cut screw threads.

SCREW-CUTTING ORNAMENTAL LATHE. This lathe has a headstock with a traversing mandrel. The mandrel is activated laterally through the bearings by means of screw guides, which provide for the pitch of the threads to be cut. Most Holzapffel lathes were of the screw-cutting type.

Screw threads can also be cut on the simple ornamental lathe by adding a spiral apparatus, which consists of a gear train that connects the mandrel to the end of the slide rest. It should be noted that the pitch of the Holtzapffel thread on the nose of the mandrel and in the chucks that fit on the nose are not of the standard pitch used today. Rather, pitch is based on the Strasbourg inch of Holtzapffel's day, 9.45 threads to the inch. The set of wheels that make up the spiral apparatus contains wheels that provide 9.45 threads to the inch.

ROSE ENGINE LATHE. The rose engine lathe exemplifies the epitome of the ornamental-lathe-making skill. The following definition of rose engine is taken from volume 30 of the *Cyclopaedia* of Abraham Rees, London, 1819:

> Rose Engine, Rose Lathe, or Figure Lathe, in the Mechanic Arts, is a machine used for turning any articles in wood, ivory or metal, in the same manner as a common lathe, but it has additional parts, by which the surface of the subject which has been turned, can afterwards be engraved with a great variety of patterns of curved lines, which, in general, are denominated from the French rosette, from a slight general resemblance which they have to a full-blown rose, and hence the machine is called a rose engine.
>
> The Rose Engine Lathe (as we prefer to name it), therefore, is a tool having the main characteristics of an ordinary lathe and in addition a device or devices for reciprocating the headstock both to and from the operator as well as longitudinally in line with the bed of the lathe, either one movement at a time or both together.

The British Science Museum has records of rose engine lathes made as early as 1740 by a German shop. Hulot & Fils made one in 1768, and another French rose engine was made in about 1800. Holtzapffel made improved versions of the rose engines, but only twenty were made by his company (figs. 14 and 15).

Norman Tweddle owned a Holtzapffel rose engine lathe and used it extensively, with great skill (fig. 16). A note in the S.O.T. *Bulletin* stated that he produced work "deeply cut and in appearance very much akin to the daintiest flowers, the rose, the violet, the daisy, the four leaf clover, the shamrock, the fine tracery of maidenfern and even the butterfly which designs, repeated around the edge of a plate, made a perfect enrichment of the pattern."

Tweddle's rose engine lathe was Holtzapffel #16, made in 1797. At the time of this writing, it is in the possession of Geoffrey Roberts of Roxborough, Massachusetts. The last true rose engine made by Holtzapffel is in the possession of

Fig. 14. The rose engine lathe made by Holtzapffel and now in the possession of Warren Greene Ogden, Jr., is possibly the finest rose engine lathe ever made. Mr. Ogden has made a lifetime study of the Holtzapffel Company and its lathes.

Fig. 15. (*Right.*) The Holtzapffel cabinet of parts and fixtures for the rose engine lathe. (*Collection of Warren Greene Ogden, Jr.*)

Fig. 16. A pattern cut on the rose engine lathe by the late Norman Tweddle, an expert on the ornamental lathe.

Warren Green Ogden, Jr., of North Andover, Massachussetts. It is reputed to be the finest such lathe ever made by Holtzapffel. It is #1636 and was first sold on December 20, 1838.

OTHER ORNAMENTAL LATHES. Other types of ornamental-turning lathes were rarely made, mostly on special order. One was the medallion-copying lathe, which copied in bas-relief a previously made master medallion that served as a template. Others were based on the lathes described above but were fitted with special apparatuses for special purposes. Some of these were made on order, but most were made experimentally by machinists and toolmakers and were never offered in the marketplace. Those that still exist are collector's items.

Basic Attachments to the Ornamental Lathe

No single attachment converts an ordinary lathe into an ornamental lathe, and it is difficult to discuss them in any sensible order of priority. The slide rest is probably the most important in terms of being able to cut even the most basic ornamental patterns, however, and so it is discussed first.

SLIDE REST. All metalworking lathes have slide rests, while few ordinary woodworking lathes have them. The slide rest, a device that fits on the lathe bearers, can be moved laterally from left to right to any desired position. It supports the cutting tools at the desired angle to the work. The slide rest can move the cutting tools back and forth across the bearers at any angle up to 90 degrees to the bearers. The tool rest on an ordinary woodworking lathe is merely a stationary metal rest on which the cutting tool or chisel is supported.

The slide rest carries a receptacle that holds the cutters. This receptacle or tool box, as it is commonly known, is activated by a lead screw, which extends the working length of the slide rest. This arrangement permits the slide rest to be maintained in a fixed position on the bearers while the cutter is moved laterally along the length of the lead screw. These movements are basic to all metalworking lathes.

The slide rest on the ornamental lathe is similar in principle to that of the metalworking lathe except that it is lighter in construction and is more flexible. It is not called on to do the heavy cutting that is common to metalworking and therefore can be adjusted to lighter types of work that are performed in the delicate cutting operations of ornamental turnery. My Holtzapffel lathe is equipped with both the metalworking and the lighter ornamental-turning slide rests. They are interchangeable.

The ornamental slide rest has a much greater throw—the distance the cutter can be moved from left to right by the lead screw without moving the rest—and this contributes to its greater flexibility. An elevating screw in the base of the slide rest allows for easy adjustment so that the cutter aligns with the exact height of center of the work, a very important matter in ornamental turnery.

A number of variations to the basic slide rest on the ornamental lathe have been designed and made by mechanically minded turners as well as the lathemakers themselves. Among these is the oval slide rest described by S. G. Askey in the February 1950 issue of the *Bulletin of the Society of Ornamental Turners*; only three oval slide rests are known to have been made. The spherical slide rest is another variation; the Reverend C. C. Ellison's private notebook states that he bought one in 1884 for £70. These slide rest variations are among the more exotic attachments to the ornamental lathe. The average turner will never need one.

INDEXING PLATE. The indexing plate is not really an attachment to the lathe but is an extension of the headstock pulley. The headstock contains a step pulley, the face of which contains several rows of evenly spaced holes that accept an index pin. An indexing plate makes it possible to fix the work in position so that it does not move while the whirling cutter produces the pattern. Moving the index pin to the next or desired hole will rotate and hold the work a fixed dis-

FIG. 17. The metalworking slide rest for the Holtzapffel lathe #2023. This rest was actually made by the Britannia Company and added to the Holtzapffel lathe, probably in the late nineteenth century. The metalworking slide rest is rarely used in ornamental turnery.

tance from the previous position. This may be repeated all the way around the work to produce an evenly spaced pattern. An indexing plate is illustrated in figure 18.

Different lathes have different numbers of rows of holes in the indexing plate. My Holtzapffel lathe has four rows of 96, 112, 144, and 360 holes. A fifth row of 120 holes is sometimes desirable. Some lathes, such as the Evans, have so many rows of holes that it is easy to put the pin in the wrong hole, sometimes with devastating results to the cutting pattern. Using an indexing plate requires close attention and patience in placing the pin.

The shank of the pin for the indexing plate has a vernier screw in the stem or at the base that allows for infinite adjustments in the division of the pattern.

OVERHEAD DRIVE. The most obvious attachment to the ornamental lathe is the overhead drive, which powers cutting tools held in the slide rest and enables them to turn, as shown in figure 9. This overhead drive is not found on modern metalworking or woodworking lathes but is peculiar to the ornamental lathe.

Historically the overhead drive has undergone a number of variations. Lathes made before the nineteenth century did not have an overhead drive, as it was a Holtzapffel development to replace former methods of activating the cutters. The motive power for the overhead originally came from a secondary belt running from the treadle and flywheel under the lathe. This belt turned the cylinder on the overhead, which had another belt that extended to the receptacle holding the cutter. The second belt could move across

Fig. 18. The indexing plate on my lathe. Figure 20 shows the index pin engaged in a hole to hold the work stationary in a predetermined position. (The oil cup is modern, made necessary by the motorization of the lathe.)

the overhead cylinder to follow the position of the cutter in the receptacle so that the cutter could move laterally. Tension on the belt that runs to the tool box has taken a number of different forms. A bow-and-spring arrangement was used on early overhead drives, but this proved to be inconvenient and was replaced with an arm with a counterbalance weight, which has greater flexibility.

My lathe, made in 1853, evidently originally had an overhead bow-and-spring arrangement on vertical posts rising from the frame of the lathe, since the indentations on the leg supports still show. At some point, some owner replaced it with a single post rising from the left end of the lathe frame. This post supports a cross member in a swinging position, counterbalanced with an iron ball on the outside end of the member. The inside end of the member holds permanently graphited wheels, which carry the belt from a motor mounted on the upright post to the cutting frame in the slide rest. This has proven to be a most convenient arrangement with easy positioning of the overhead belt to meet any desired position of the cutter. Ordinary dental belts can be used quite successfully in the overhead drive.

Fig. 19. Just a few of the scores of attachments for the ornamental lathe.

CHUCKS. The active ornamental turner never has quite enough chucks. Thirty-four different chucks came with my lathe, and at the time, there seemed to be enough of them to serve several workers. They all seem to get used sooner or later, however. Among the most frequently used chucks are the faceplate chuck and cup chuck, which are available in a range of sizes and are made of heavy brass. Probably the most used of all the chucks that came with my lathe is a four-inch Cushman 3-jaw self-centering chuck adapted to fit the mandrel nose. It has both inside and outside jaws. Other useful chucks include eccentric chucks, driver chucks in two sizes, metal-holding as well as wood and ivory-holding chucks, spiral chucks, and elliptical chucks.

Some very complicated chucks, particularly the elliptical, geometric, epicyclodial, and rose engine chucks, have been designed and made for the lathe. These may be considered exotic attachments, made either to satisfy the creative instincts of a master toolmaker or to resolve a particular turning problem and probably never used again.

FIG. 20. The eccentric chuck fastens onto the mandrel nose and carries an auxiliary nose that can be moved out of center into eccentric positions. This particular attachment was made by James Munro to fit Holtzapffel lathe #2023.

CUTTING FRAMES. Cutting frames hold the cutters in the receptacle or tool box on the slide rest. The tool box contains a groove, into which the 9/16-inch-square shanks of the cutting frames fit. Some shanks are solid while others are drilled longitudinally to accept a revolving stem. Figure 21 shows an eccentric cutting frame, which consists of a shank with a revolving stem and a T that, in turn, contains a lead screw to position the cutter at varying radii from center. This device cuts circles of predetermined radii on the work and can implement many interesting designs. The barleycorn pattern is made with this cutting frame.

Basic ornamental turnery requires only three cutting frames to do a wide range of work: the vertical, the horizontal, and the drill (fig. 21, *e*, *b*, and *c*). Both horizontal and vertical cutting can be done with the universal, figure 21*a*, which can

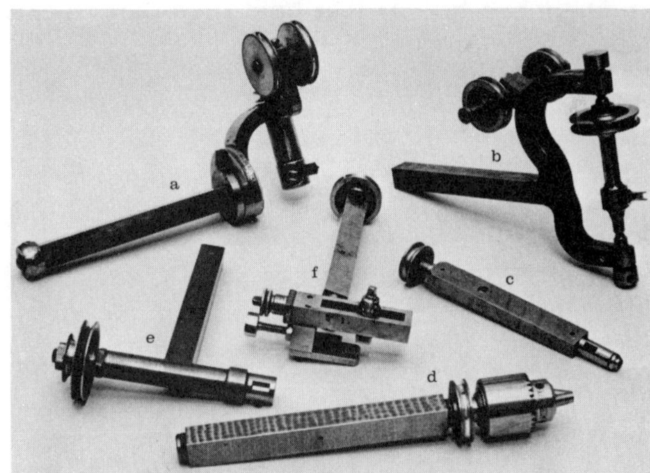

FIG. 21. Cutting frames, which hold the cutters: *a* is universal, swinging through 90 degrees of position both sides of center; *b* is for horizontal cutting; *c* holds drills that cut on center; *d* is a modern version of the drill with a 3-jaw chuck; *e* is for vertical cutting (there is another with a short T arm; and *f* is an eccentric in which the cutter may be moved out from center to cut circles. (*Photo by John Kelsey for* Fine Woodworking)

be positioned at all angles through 180 degrees—90 degrees to the right and left.

Other cutting frames are the side cutter and, for the more exotically inclined, the elliptical, the epicyclodial, and the rose cutting frames. These are not necessary for the average ornamental turner. All are described in the Holtzapffel volumes.

SPIRAL APPARATUS. The spiral apparatus consists primarily of three parts: a special chuck, a banjo arm, and a series of gear wheels, plus special fittings for its assembly.

The spiral chuck screws onto the inner end of the headstock shaft, or mandrel nose, and has a secondary nose that carries the work. It also carries the first of the train of gear wheels. It has a special indexing wheel with a spring-loaded arm with four teeth that engage the ninety-six teeth on the chuck's indexing wheel, thus taking the place of the ninety-six holes in the main index on the face of the pulley. This limits the indexing of spirally turned pieces to divisions of ninety-six.

The banjo arm is a heavy brass arm that fits on the face of the headstock surrounding the base of the spiral chuck. The arm extends out toward the turner and has a slot that accommodates the fittings that hold the other wheels in the gear train. It is fastened solidly to the headstock but can be moved up and down to accommodate different sizes of gear wheels on the arm. In use it is locked in position.

The spiral set consists of sixteen brass wheels of increasing sizes. The smallest wheel has 15 teeth; the largest has 144. The train itself can consist of either three or four wheels, with a fifth idler wheel added when the direction of the spiral must be reversed. With the various gear wheel settings, it is possible to cut spirals or threads with as many as forty turns to one inch or as few as one turn or thread in approximately seven inches—the slow spirals used to decorate vases, candlesticks, and the like.

CURVILINEAR APPARATUS. This apparatus permits the decoration of curved surfaces, either straight or spiraled. It consists of posts mounted on either end of the slide rest that support a bridge that spans the entire length of the rest. Mounted on this bridge, a template governs the route to be taken by the cutter to decorate the curved surface.

To do this the tool box is freed from the collar that fixes it to its control screw. This permits it to move freely in and out, guided by a handle connecting the tool box with the frame of the slide rest. Mounted in the top of the tool box is

Fig. 22. Attachments for the spiral setup. The gear train connects the mandrel nose to the end of the slide rest to provide a range of spiral pitches from forty turns in one inch to one turn in seven inches. This is one of the most frequently used of the lathe's attachments.

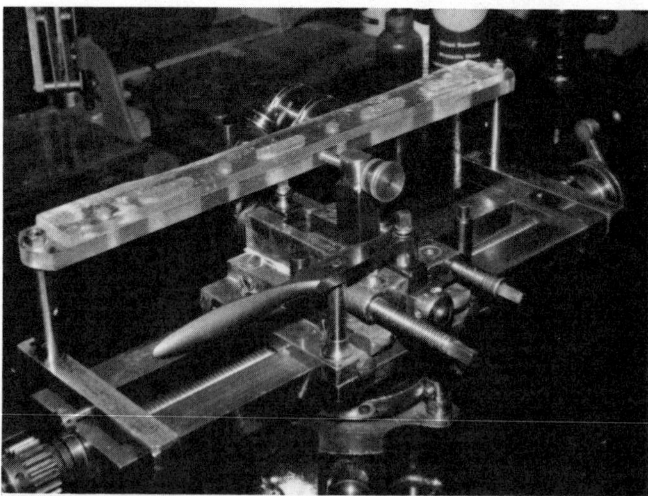

Fig. 23. The spiral and curvilinear setups mounted on the lathe, showing the bridge that guides the cutter along the work.

Fig. 24. Another view of the spiral and curvilinear setups showing the template fastened to the bridge to guide the cutter around the curvature of the work.

a rubber, which has a shaped end that rubs against the curved edge of the template mounted on the bridge. Thus, the cutter follows the exact curvature of the template.

STEADYREST. On many occasions it is necessary to support the outer or tailstock end of the work when the tailstock cannot be used to do so. This happens particularly when end-boring or hollowing is being done. Holtzapffel made a boring collar which contained a series of holes spaced around the circumference to permit the entrance of a boring piece. However the largest hole in this piece is 1¾ inches in diameter, and in many cases a hole is needed to accept a greater bore in the piece being worked. The Holtzapffel boring collar does not provide this. Furthermore, the adjustment of that boring collar to the exact height of center is sometimes quite difficult.

The answer to this problem is a steadyrest. It is an octagonal metal frame—placed over the work piece after the surface is trued—with four pins spaced equidistantly around the circumference of the frame. The pins slide through holes in the frame until they rub the surface of the work, which is centered within and protrudes

Fig. 25. A modern steadyrest, which supports the middle or outer end of the workpiece while end boring or hollowing out the inside of a bowl. The original Holtzapffel boring attachment had seven fixed holes ranging from ⅜ to 1¾ inches in diameter, which were insufficient in size to support larger workpieces. (This steadyrest was acquired from Model Engineering Services, 6 Kennet Vale, Brockwell, Chesterfield, Derbyshire, England.)

slightly from the frame. The pins are then locked in position. The inside diameter of the frame is slightly greater than five inches, large enough to accommodate most turnings. The steadyrest is very easy to mount. First, the workpiece, supported by the tailstock, is plain-turned to create a true surface that will accept the ends of the pins. With the work turning slowly and the tailstock still in place, the pins are dropped into position on the surface of the workpiece. The pins are locked and the tailstock is removed; the piece is now supported accurately on a centerline.

Cutting Tools

One of the most troublesome problems facing the ornamental turner is compiling a reasonably complete set of cutters. To illustrate: one turner in recent years bought a lathe only to discover that exactly one cutter came with it. On the other hand, a well-equipped lathe will have a hundred or more cutters.

TYPES OF CUTTERS. The cutters are tiny chisel-shaped tools that actually cut the patterns. They fit in the tool box receptacle in the slide rest and are not hand-held as are chisels in plain turnery. They come in different sizes, widths, lengths, and contours. Among the infinite variety of cutters are square-end, convex, concave, single-angle, double-angle, quarter-hollow, half-round (in matched pairs), and double-concave cutters, drills in most of these shapes plus drills in a stairstep pattern, thread cutters (in matched pairs), side cutters in various contours, plus many others. Each of the shapes comes in as many as twelve or fifteen widths. You can see how quickly a

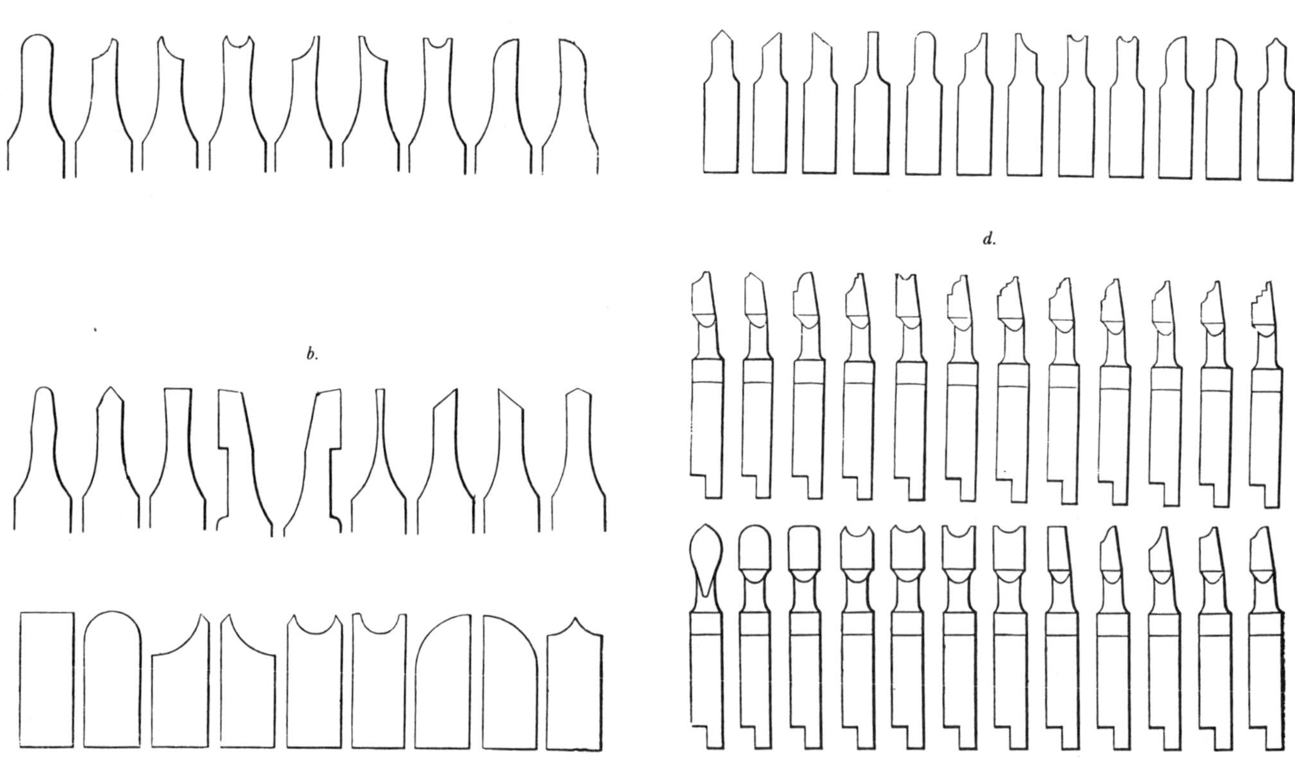

FIG. 26. Typical sizes and shapes of cutters for the Holtzapffel lathe: *a*, four-inch-long fixed-position cutters; *b*, cutters used in the horizontal, vertical, and universal cutting frames; *c*, cutters used in the eccentric cutting frame; *d*, cutters used in the drill cutting frame.

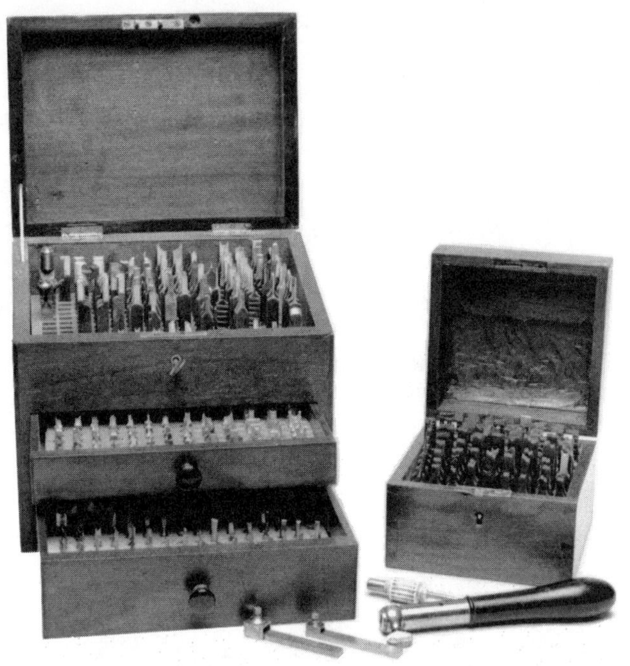

Fig. 27. Original Holtzapffel mahogany boxes of cutters for the ornamental lathe, containing more than 650 cutters with no duplicates.

"complete" set of cutters can add up to many hundreds. In addition I have a small mahogany box fitted with fifty-four specially ground cutters in molding shapes. Who cut them? I don't know, but the box is obviously from the nineteenth century.

Holtzapffel made special mahogony boxes with drawers and hinged covers to hold the cutters. The boxes themselves are collector's items. Along with the cutters came special handles into which the cutter can be inserted and fastened with a brass thumbscrew. These handles allowed the turner to transform ornamental-turning tools into plain-cutting tools. Holtzapffel even included a special pair of tweezers for plucking the cutters from their nests.

SHARPENING THE CUTTERS. When I obtained my lathe, the friend in England who found it for me suggested, "Now that you have your lathe, I recommend two things. First, retire, and second, teach your wife to sharpen the cutters." Only the first has come about; the second is still a somewhat onerous task.

The standard sharpening device for ornamental cutters is called a goniostat and is illustrated in figure 28. It is essentially a three-legged device of which the cutter forms the third leg. The two fixed legs move on a fixed surface while the cutter moves on a sharpening stone or a metal plate charged with lapping compound. The tool leg can be adjusted through angles of 54 degrees right and left for double-angled cutters and can be adjusted through 70 degrees in the other direction to accommodate the rake of the tool. The latter is normally fixed to a predetermined rake angle and not moved.

Another sharpening device, designed by the

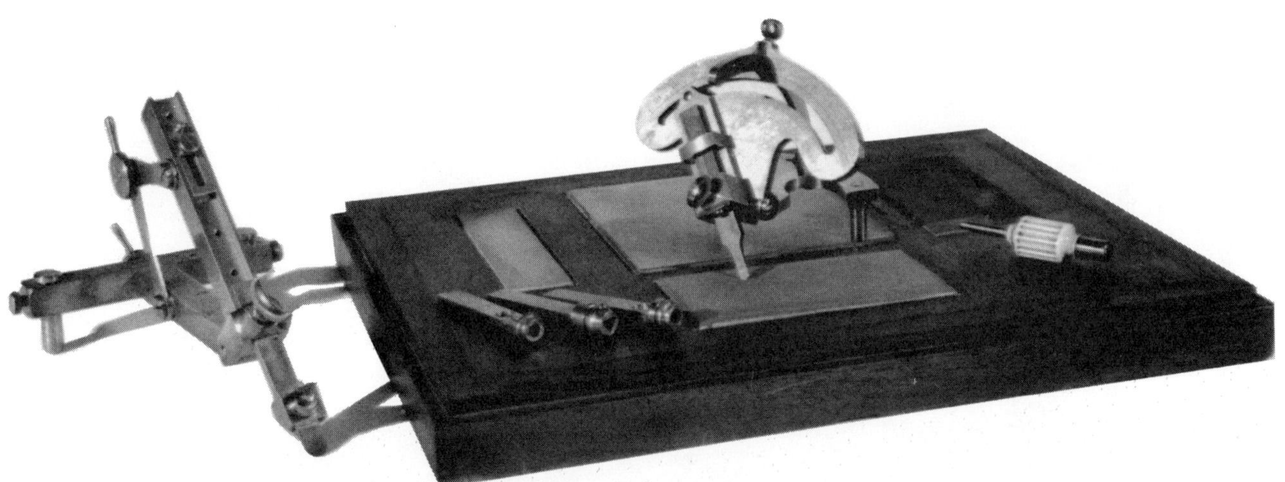

Fig. 28. Holtzapffel goniostat for sharpening flat and angled cutters. At left is a modern goniostat for sharpening convex cutters, developed by the Reverend G. A. Grace of the Society of Ornamental Turners.

Reverend G. A. Grace, is similar in principle but takes a round-ended tool and rocks it back and forth on the stone, with excellent results.

Because delicate turned cuts cannot be finished with sandpaper, the cuts normally create the finished surface. This means the cutters must have razor-sharp edges. This is normally accomplished by the use of hard Arkansas stones.

Concave tools are sharpened with an appropriately sized cone. The cones come in sets of progressive sizes to accommodate all the diameters of the concave shapes. A device that fits on the slide rest of the lathe accepts and revolves these cones, which are charged with lapping compound for sharpening. There are two sets of cones: brass for first sharpening and iron for the final grind. Lapping compounds range from 120 to 1000 grits.

Sometimes it becomes necessary to pause in the middle of a cutting operation to resharpen the cutter. It is important to very carefully reposition the cutter in the cutting frame after sharpening it. Otherwise the design will be moved slightly out of position; this mistake may be surprisingly visible in the finished piece.

FIVE

Preparing for Ornamental Turnery

If a piece of turnery is to be embellished or ornamented, you obviously must begin by plain-turning the workpiece as completely as necessary. If you are already a woodworker, chances are that you are reasonably competent in plain turnery. If you have no woodworking experience, you will have to become competent in plain turnery before you can ornament. Because ornamental turnery can require very close tolerances at times, the turner must be able to plain-turn his initial piece with such tolerances, particularly if the piece requires assembly, such as a stemmed chalice with a cover and finial or candlesticks with different woods composing the cup, stem, and base.

This does not mean that the turner must match the tolerances required of an engineer or toolmaker, who may work with thousandths, even ten-thousandths, of an inch. Wood is not stable enough to hold the tolerances commonly used by the toolmaker; it can change shape overnight on the lathe to a much greater degree than would be acceptable to the toolmaker. But, just because the exacting tolerances of engineering are not required does not mean you should not try to achieve the most precise tolerances you can in plain-turning. Do not rely on glue to cover your mistakes. Common glues are not crack fillers and will not by themselves correct a gap caused by poor fitting of the pieces to be assembled. You should be competent enough in plain turnery to avoid gaps and discrepancies in curved surfaces.

Any competent woodworking school can teach you all you need to know to become a fairly competent turner of plain objects. The differences between slicing and scraping the wood and when each procedure should be used; the correct use of the gouge and skew chisels; the use of the parting tool and other special chisels; and the degree to which the tool should eliminate the use of sandpaper in finishing the surface—all this and more can be learned.

The beginner will be making a grievous mistake if he or she thinks that such study of plain turnery is not necessary. It means that at some point, not far distant, the ornamental turner will have to back up and correct the mistake, learning the necessary basics of turnery.

Concept and Design

Anyone competent in plain turnery can mount a piece of wood on the lathe, shape it into a cylinder, and decorate it with an ornamental pattern, but all this produces is a decorated piece of wood. This exercise is fine for beginners practicing the cutting patterns; indeed it is a necessary learning exercise. But rather soon you will want to produce something that is functional as well as decorative, something that could be classified as decorative art.

A work of art or craft will either attract or repel a viewer at first glance. If the object is useful or simply lovely to look at, it will attract. If, on the other hand, the object has no meaning for the viewer, if it is unrecognizable or weird, it may well repel. If the craftsman is working only for his or her own edification, this makes no difference. If he or she wants the piece to be admired or possibly sold, however, the concept or design must take into consideration the reaction of potential viewers and/or buyers.

This does not mean that the craftsman must be a second Michelangelo or even an original creative artist. Not at all! What it does mean is that he or she must first decide what impact the finished piece is to have on viewers and then create a piece that fills those requirements. If it is to be a bowl or vase, a candlestick or chalice, the turner can follow one of two possible paths: create the piece originally using imagination and knowledge of traditional design, or check the stores and read the ads in the magazines and then base the design on observed pieces. This does not mean that the observed piece should be copied, but it can serve as a point of departure, and can be adapted to fit the requirements or niceties of ornamental turnery or the piece of material available for it.

Of course, it is much nicer if the piece is an original design made by the turner. Many wood turners, however, feel that they are incapable of such original creation. Do not be intimidated by design. Look around at what has been done to get ideas and shape them as you wish. To some degree, every artist through the centuries has done this, for all people are influenced to some extent by the objects that surround them.

Books and schools can teach you basic design principles such as the golden proportion, dynamic symmetry, and the Carbonacci curves. As you continue with ornamental turnery, you may want to learn more about design. But you can create pleasing objects without this formal knowledge. Above all keep in mind that the viewer will judge the piece by his or her own reaction to it, not necessarily by your intentions.

The material from which the piece is made also plays a role in design. If the piece has an exciting concept but is made from very plain or even unattractive material, it may not attract the viewer. An unusual or particularly beautiful material will enhance an attractive design. A vase, attractive in shape and design but made of some ordinary wood with no color or grain pattern, will have less appeal than if that same vase were made of Brazilian kingwood, which is attractive in itself.

The third factor involved in design is the technique used to realize the piece. Here is where ornamental turnery has unique appeal. A handsome design made from a wonderful material can be given additional appeal by ornamenting or embellishing its surface. Decorative patterns will draw the eye by creating a texture on an otherwise plain surface.

So, as you design ornamental turnings, you should consider three factors: the shape of the object itself, the material from which it is made, and the techniques used in finishing or decorating the piece. Viewers will usually evaluate the piece in that order. Seldom does anyone say, "That is beautiful because it is an example of ornamental turnery," but after their initial reaction, they may well say, "But how in the world did you accomplish that pattern on the surface?" All this is a matter of advance planning on the part of the turner.

Woodworking Technique

Just as plain turnery serves as a foundation for ornamental turnery, so all forms of turnery are based on elementary woodworking. Knowledge of woodworking is not only desirable but is a practical necessity. For example, if a piece such as a candlestick or chalice is in several parts and must be assembled, you must understand joinery in order to complete it.

It has been said that basic woodworking requires only four skills: measuring, marking, cut-

Fig. 29. A pair of boudoir lamp bases made from Nigerian ebony, African stinkwood, and African blackwood, with collars of ivory. Designed, crafted, and photographed by the author. (*Collection of Dr. and Mrs. Miles Taylor*)

ting, and joining. An over-simplification? Of course, but when you begin woodworking, you find out that there is a basic truth in the statement, and it applies to ornamental turnery as well. Look at the base of the chalice in plate 3. Its construction was essentially an exercise in woodworking. Because there was not enough purpleheart wood to make it from a single piece of stock, four pieces had to be cut, beveled, and joined at the corners to make the cube. This involves measuring the pieces, marking them for exact cutting, cutting them at the correct angle, and joining them so the corners met with complete accuracy. Then the piece was joined to the ebony base, and the molding was cut on the router and carefully fitted at the corners preparatory to gluing up. The inlays were fluted on the ornamental lathe and glued into the recesses that were in turn cut on the ornamental lathe. Even if such a piece is made with a round base, which can be made on the lathe, you must still understand the fundamentals of woodworking: measuring, marking, cutting, and joinery.

AUXILIARY WOODWORKING TOOLS. If the beginning ornamental turner already has an existing workshop, he will probably have all the auxiliary tools he needs. For the most part these tools are commonly found in any woodworking shop. To start, a band saw and a disc sander are necessary. To prepare the raw stock for making the plain-turned piece, the band saw is invaluable for both straight cutting and cutting curves. The disc sander is used to surface the piece before attaching it to the chuck with double-faced tape. Many more uses will be found for these two tools. If you do cabinetwork, you probably have a circular saw, but this tool is used infrequently in ornamental turnery and is not a substitute for the band saw. A drill is also necessary, although horizontal drilling can be done on the ornamental lathe. Ordinary bench tools—rules, knives, wrenches, screwdrivers, and the like—are necessary for ornamental turning as well.

In addition to tools, you must also give some thought to your workshop space. I work in a small space measuring about 12 by 18 feet. It does not have room for individual set pieces such as the circular saw, drill, sander, joiner, and the other tools necessary for a cabinet shop. I therefore use a small all-purpose machine that incorporates the circular saw, lathe, disc sander, and drill, which operates with a single motor. Attachments that can be operated from the same motor include the band saw, belt sander, jigsaw, and jointer. In such a small space, this is the only possible solution. I have been working in this way for more than twenty-five years with no trouble. Having studied cabinetmaking for more than ten years prior to obtaining my ornamental lathe, I am tolerably efficient in plain turnery and basic cabinetmaking. When I have a piece of cabinetwork that cannot be handled in my small shop, which happens seldom, I turn to a professional woodworking shop for assistance. It works out quite well.

SIX

Chucking and Mounting the Workpiece

ANYONE competent in plain turnery is acquainted with the problems of chucking, fastening the material on the lathe, both in cylindrical and faceplate turning. The oldest and possibly the most common method is that of using screws to fasten the work to a faceplate or to a piece of scrap wood. Ornamental turnery, however, presents problems not normally encountered in plain turnery, occasioned by the many operations possible on the ornamental lathe but not possible on plain lathes.

Unusual chucking problems in ornamental turnery occur when:

1. The work must be removed from the lathe and remounted, possibly several times, without losing the centerline. For example, the turner may want to remove a plain-turned vase from the lathe to work with a design in the round—which is not possible to do using a one-dimensional drawing. When the piece is remounted to check the indexing, he/she may find that the wood has changed its shape overnight and relocating the centerline may be difficult.
2. The work must be reversed end for end and reworked without losing the centerline, possibly reversing it a second time to bring it back again to the first position for final working. For example, a particular piece may require the drilling of dowel holes before mounting—in a plug-and-recess cut—into the scrap-wood chuck.
3. The work must be cut into two or more parts after being initially worked, each of the two parts further worked separately, with the second part requiring a second chucking, and then reassembled without losing the centerline. An ivory or other inset band that requires affixing, and therefore shoulders that must be cut in the work both above and below the insert, may cause problems in relocating the centerline.

Methods of Chucking

Before examining the mechanics of mounting the work to resolve the above problems, the various basic methods of chucking must be reviewed. Some of them are essential to ornamental turnery while others, although not absolutely essential, have proven very convenient. I have used the following methods for several years.

PAPER SANDWICH. The paper sandwich is used frequently in both plain and ornamental turnery, usually for faceplate turnings. In many cases the revolving tools must run off the edge of the wood and the paper sandwich is used to prevent the tool from hitting the metal. The waste block and the work are both carefully and accurately faced (very easy to do with the ornamental slide rest and a fixed cutter) and are glued together with a piece of moderately heavy paper—craft wrapping paper is excellent—between them. They are clamped and allowed to dry. The waste wood is then screwed into the faceplate, which in turn is mounted on the headstock center. After the turning is complete, the edge of a firmer chisel is placed at the edge of the paper and lightly tapped. The paper will neatly split in two, half on the waste block, half on the work, thus freeing the workpiece from the waste block. The only remaining task is to remove the residual glue and paper from the workpiece, an operation that sometimes seems a nuisance, especially when considered in the light of the next method.

DOUBLE-FACED-TAPE METHOD. This is one of the most efficient methods for faceplate chucking. The work is attached directly to a faceplate or to a piece of waste wood that is fastened to the faceplate with a piece of double-faced adhesive tape.

In order to achieve the proper degree of adhesion, the work and the faceplate must be clamped, but only for a minute or two. The clamp compresses the tape into both the mount and the workpiece, and under normal circumstances the workpiece will not separate from the mount after turning begins. A more likely problem is not being able to separate the two after the turning is completed. A few drops of solvent dripped into the edges of the tape or in the holes on the back of the faceplate will loosen the bond. You can then insert a spatula to remove the work. The solvent is also used to clean the residue of tape from the work and the chuck.

Three kinds of double-faced tape are on the market, paper-, plastic-, and cloth-based. *Paper-based tape* is commonly used for fastening carpet to the floor but is totally unsatisfactory for mounting. The paper base is so thin that it will not adapt to or equalize the minute (and sometimes not so minute) inequalities in the surfaces of the work and the mount, resulting in patchy adhesion. *Plastic-based tape* may work, but it forms such a strong bond that it is difficult if not almost impossible to loosen the work from the chuck after turning. *Cloth-based tape* is the tape to use. The cloth base is normally thick enough and has sufficient compressibility to adapt to the inequalities of surface and forms a strong but not inseparable bond.

PLUG-AND-RECESS METHOD. Whereas the tape and paper sandwiches are ideal for faceplate work, they cannot be used for cylindrical turning between centers because they do not provide sufficient rigidity to the piece for lateral tool pressure. Here the plug-and-recess method is most satisfactory. This method consists of cutting a plug or circular protrusion onto the end of the work and gluing it into a shallow recess in the chuck. The recess may be cut into a piece of wood previously driven into a cup chuck or into a piece of waste wood mounted on a faceplate to protect the piece from revolving cutters. Such a mount requires the use of the tailstock center, first to center the work when gluing into the recess and second to support the outer end while turning. Sometimes a workpiece that is being end-bored or that is relatively short in relation to its breadth or thickness, such as for a bowl, will be worked without the tailstock, sometimes supported by a steadyrest, but normally the tailstock center should be used.

Most Holtzapffel lathes come equipped with several sizes of heavy brass cup chucks, and these are normally used for the plug-and-recess method. It is a simple matter to select a cup chuck and drive a wooden plug into it. Usually friction alone will hold the wood plug in the cup chuck, but if you have any doubts, a drop or two of ordinary white glue or small pieces of double-faced tape placed between the plug and the edge of the cup will prevent slippage during turning. The cup chucks are very slightly tapered inside and a plug, cut to start in the cut, can be driven further with a strong clamp. (Always be sure that the chuck is protected from possible damage or marking from the clamp by using a piece of scrap wood between the two.) It is then a simple matter to face off the wooden insert and cut a ¼-inch-deep recess in it to receive the plug cut into the end of the workpiece. Use the tailstock for centering.

Tailstock Accessories

Several accessories to the tailstock are necessary in order to mount the workpiece on it. These accessories may, in some cases, have to be specially made. These accessories are especially handy for centering work, for reversing the workpiece end for end on the lathe, for interchanging workpieces between lathes while work is in progress, for gluing up workpieces on the lathe without losing the center, and for center-boring workpieces.

TAILSTOCK ARBOR. The first tailstock accessory is an insert tapered to fit the recess in the tailstock (fig. 30). It will have an arbor of about ⅝-inch diameter on the protruding end, to hold additional accessories. It is flattened on one side to accept a setscrew from the accessory being mounted.

SELF-CENTERING CHUCK. A 3-jaw self-centering chuck to fit the tailstock arbor is a useful accessory. I use them probably more than any other single chuck. All end-boring bits are clamped in the tailstock with the 3-jaw chuck.

HOLTZAPFFEL LATHE NOSE CHUCK. This is a chuck with a Holtzapffel nose and a ⅝-inch aperture that fits on the tapered tailstock arbor. This very useful chuck will probably have to be made to order with a Holtzapffel thread (9.45 threads per inch). It accepts reversed work already mounted on another chuck.

Mounting a Workpiece to Turn Between Centers

Assume that you want to mount a cylindrical workpiece, since it is the most adaptable to the vertical and horizontal cutting frames. The following steps should be followed for mounting the workpiece.

Step one: Select the workpiece and center each end. Select a piece of wood that will produce a cylinder of suitable diameter, approximately 6 to 8 inches long. Find the exact center on both ends and punch a small guide mark in each. A center-finding device is helpful in quickly and accurately locating the centers.

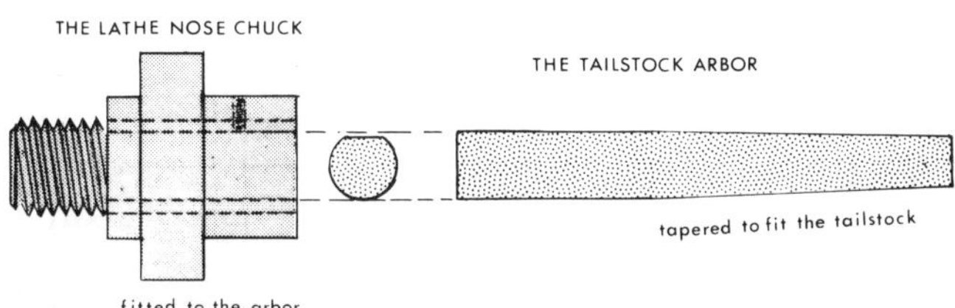

Fig. 30. A modern tailstock arbor, tapered to fit in the Holtzapffel tailstock and accept fittings such as the auxiliary lathe, nose chuck as shown, and other attachments.

Step two: Rough out for chucking. Assuming the piece of wood is in the square, knock off the outer corners about one inch in from one end with a band saw, making that end roughly into an octagon.

Step three: Mount in a 3-jaw chuck. Place the 3-jaw self-centering chuck on the mandrel nose. Mount the octagonal end of the wood loosely in the chuck.

Step four: Place the countersink bit in the tailstock. Place the tailstock arbor with a ⅝-inch 3-jaw chuck attached to it in the tailstock. Mount a countersink bit in the 3-jaw chuck.

Step five: Countersink a center hole in the wood. Advance the tailstock until the point of the countersink bit meets the center of the end of the wood. Tighten both the tailstock and the chuck on the octagonal end of the wood. Now, with the work held solid and centered at the tailstock end, start the lathe in slow motion. Run it just long enough to advance the countersink bit into the wood to create a shallow hole to receive the tailstock center.

Step six: Replace the countersink bit with a

live center. Remove the complete auxiliary assembly from the tailstock and substitute the regular tailstock center (preferably a live center). Advance the tailstock to the shallow hole just cut in the wood and lock it in place. Plain-turn the tailstock end, an inch or two only from the end, into cylindrical form. (Remember, the octagonal end in the 3-jaw chuck is not truly centered.)

Step seven: Cut a projection for plug-and-recess mounting. A plug ¼ or, at the most, ⅜ inch long and somewhat less in diameter than the diameter of the cylinder is cut on the rounded tailstock end. The reason for the smaller diameter of the plug will become apparent in later work, when a revolving cutter must go off the end of the work.

Step eight: Reverse the work end for end. Remove the work from the lathe and reverse it end for end, clamping the trued round end with the plug into the 3-jaw chuck. Reassemble the countersink bit assembly in the tailstock and cut a tapered hole in that end to receive the regular tailstock center. Remove the countersink bit assembly and remount the tailstock center. Plain-turn the workpiece to a true cylinder.

Step nine: Mount the work on the mandrel nose. Remove the cylinder from the lathe. Mount a cup chuck with a piece of wood driven into it onto the mandrel nose. A piece of wood mounted on a faceplate may also be used, but be sure that the wood is sufficiently stable to prevent splitting during turning. Turn a recess in the wood to receive the plug cut on the end of the workpiece. Glue the plug into the recess, centering the work with the tailstock center. If the recess has been cut for a snug fit, the pressure of the tailstock center is normally sufficient to hold them together. If a quick-acting glue is used, the workpiece should be ready for further turnery in a half hour. We are now ready to consider the application of one of the Holtzapffel cutting frames.

SEVEN

Cutting Patterns

CUTTING ornamental patterns is the essence of ornamental turnery. No craftsman can live long enough to exhaust the multitude of possible cutting patterns that can be created on the ornamental lathe. With the tremendous variety of cutting frames, cutters, special chucks, and other attachments available, only a slight adjustment of any one of these elements will produce a different pattern.

In the following sections, the patterns produced by the major components of the lathe will be described in detail. Factors common to most, if not all, patterns will be described in terms of the vertical cutting frame.

Segment Cuts—The Vertical Cutting Frame

Figure 31 shows a cylindrical workpiece (end view) with a vertical cutter making a segment cut into it. In this case the cutter has been given a ¾-inch radius, and the index pin is moved six holes for each successive cut, using the 96 row of index holes. Thus, there will be sixteen segment cuts around the work (96 divided by 6 equals 16), to a depth shown in the left-hand side of figure 31. Each cut just touches the edge of the adjoining one. The measurements are determined by the diameter of the workpiece and the requirements of the design.

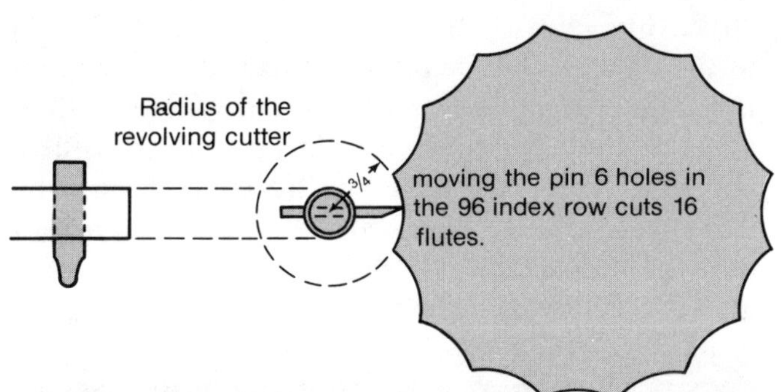

FIG. 31. This diagram shows the position of the vertical revolving cutter in relation to the work when cutting a flute one-sixteenth of the circumference of the work.

POSITIONING THE INDEX PIN. When the sixteen segment cuts have been made around the work, the result will be the first cylindrical row of segment cuts shown in figure 34, no. 112. For these cuts the index pin should be placed, successively, in holes 96, 6, 12, 18, 24, 30, 36, 42, 48, 54, 60, 66, 72, 78, 84, 90, and back again to 96—to make the next row of cuts.

POSITIONING THE CUTTER FOR SUBSEQUENT CUTS. For the second row of segment cuts, the cutter will be moved to the right (assuming you are working from left to right) the exact width of the tool. This movement can be precisely achieved with mechanical measurement using the lead screw of the slide rest.

In reality this becomes more principle than

practice unless the cutter widths are absolutely accurate to hundredths of an inch—something rarely possible with tools made over a century ago. Rather, the movement may be governed visually by advancing the cutting edge to the surface of the work and moving it over its own width, minus a very minute fraction. If done with absolute accuracy, the edges of the first, second, and successive rows of segment cuts will meet exactly with no gap or overlap between them, but this kind of accuracy is hard to achieve visually.

If the cutter is moved just a hair short of where it should be, the cuts will overlap very slightly with the first row, but this will not be noticeable except under the most careful scrutiny. If, on the other hand, the cutter is moved just a hair too far, it will leave a very thin but noticeable wall of wood between the rows.

The use of a three- or five-power magnifier attached to eyeglasses or a headband will facilitate this spacing operation. Note: The turner should always wear goggles or some type of eye protector to prevent chips, dust, and the like from getting into the eyes.

DETERMINING LENGTH. If there are to be sixteen segment cuts around the work, each cut should obviously be 1/16 of the circumference of the work, not 1/15 or 1/17. How to get the cuts started accurately?

After placing the index pin in the ninety-sixth hole, advance the depth screw so the cutter meets the work. Then retract the depth screw slightly and make a cut to that depth. Move the index pin to the sixth hole and make another cut. The ends of the two cuts will not quite meet, so retract the depth screw a little more, make the cuts again, and repeat until the ends of the segment cuts just touch. This is the correct depth of cut, and all successive segment cuts in that row can be made with the depth screw in that position.

In actual practice the cuts will be made very slightly deeper so the ends of the cuts will meet sharply regardless of any deviations in the surface of the wood. Put another way, the original cylinder may be left slightly greater in diameter than the desired finished surface to allow for a slightly greater depth of cut than if a perfect alignment of cuts were achieved in relation to a true dimension of the finished work. The latter is, theoretically the proper way to form the cuts, but it is not often achieved when the depth of cut is controlled visually.

IDENTIFYING THE CORRECT INDEX HOLE. To cut segments progressively in the first row, the index pin is moved manually to the next hole, in this case from 96 to 6 to 12 to 18, and so forth. These holes are not all marked on the indexing plate and must be located and identified visually. In each row of holes on the index face, certain holes are identified by a number impressed in the brass. In the case of the 96 row, each twelfth hole is so marked. On my lathe some previous owner scratched identifying marks at certain intermediate holes, 6, 18, 30, and so on. The purpose, of course, is to be able to locate the proper intermediate holes when working the pin around the row to guarantee even spacing of the segment cuts. Neither the indented numbers or the intermediate scratches, however, are very easy to see under the light normally available in the average shop.

This situation can easily lead to regrettable errors. The pin is accidentally inserted into the wrong hole, an improperly spaced segment cut is made, and the entire pattern is damaged if not ruined since such improperly spaced cuts are very apparent in the finished design.

I use a very simple method to avoid this potential error. With white enamel and a fine brush, I have painted a small white circle around those holes that afford a quick benchmark against which intermediate holes can be counted. On the 96 ring, the holes painted are the 96, 6, 12, 18, 24, and so forth. The same pattern was followed on the other rows of holes, selecting divisions of those rows in a similar manner. This greatly simplifies the quick identification of the correct hole for the segment cut. The white enamel does not injure the index face, is easily retouched when worn, and is easily removed if so desired.

THE SECOND AND SUBSEQUENT ROWS OF SEGMENT CUTS. After the first row of segment cuts is complete, the cutter is moved to the right, assuming you are working from left to right, its exact width minus a hair. The index pin must then be placed in a different series of index holes to provide the desired pattern. For example, the basketweave pattern, shown in figure 34, pattern *a*, no. 112, requires that the second row of cuts be made midway between those of the first row. The index pin is therefore placed in holes 3, 9, 15, 21, 27, and so on, halfway between the cuts made for the first row. This will result in the basketweave pattern. The third row of segment cuts reverts to pin placements in 96, 6, 12, 18, etc., as in the first row and each subsequent row will follow this same sequence of change.

FIG. 32. Wassail bowl, done in lignum vitae wood, modeled after the wooden vessels used in England in medieval days. This piece illustrates the wide variety of designs that are possible to create using the ornamental lathe. It also shows the tendency of some medieval turners to over-decorate their pieces. Crafted by Mr. D. C. Gall of the Society of Ornamental Turners in England. (*Reproduced by permission of the Society of Ornamental Turners*)

FLUTING STOPS. Under some circumstances, it is possible for the slide rest's lead screw to change position laterally after the cutter has been placed in position to cut a given row. The locking of the fluting stops against the tool box on the slide rest will prevent any possibility of this lateral movement. The fluting stops, of course, have to be repositioned with each lateral movement of the cutter for making subsequent rows of segment cuts. Is it a nuisance to do this? Perhaps, but it's better than having a piece of work spoiled when someone accidentally brushes against the crank of the slide rest and moves the tool box where it is not wanted. Vibration or even the weight of the crank itself can also effect such a change.

THE DIRECTION OF CUTTER REVOLUTION. Holtzapffel speaks of the vertical cutting tool being driven to cut *downward* to throw the shavings away from the surface of the slide rest. Initially, I followed this recommendation until it ruined a workpiece when the downward force of the cutter loosened the work on the screw of the mandrel nose, thereby violently stopping the cutting operation when the cutter encountered the uncut part of the work.

The workpiece, of course, screws onto the mandrel in a clockwise direction, and there is no steady pin to lock it in place. Thus, the workpiece

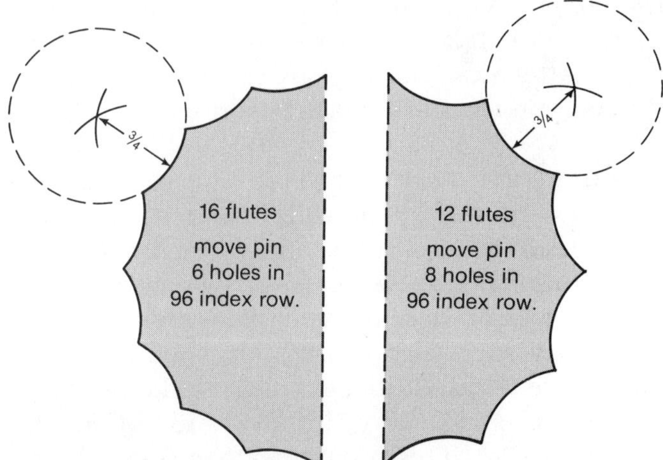

FIG. 33. This diagram shows the position of the cutter in relation to the work when cutting variable numbers of flutes.

can be positively fixed only by friction between the face of the mandrel and the chuck. If the chuck has to be forced into a very tight fit, it becomes more difficult to remove from the mandrel and, even then, the possibility of loosening remains.

WEAR ON THE CUTTING FRAME. The pull of the overhead belt on the end of the T of the cutting frame plus the possibility of wear in the fitting of the shank of the frame as it seats in the groove of the tool box may result in a slight tilt of the cutter in its vertical cutting path. This puts each successive cut out of true, at a slight angle to the preceding and following cuts, and can create an inaccurate pattern that is easily visible. If this happens, place a shim in the groove of the tool box to compensate for the tilt.

My lathe has two vertical cutting frames, one with a short T and one with a long T. Both unfortunately show some such wear, which is not particularly noticeable on the short T but is definitely noticeable on the long T. Although this is correctable, as explained above, I have found using the universal cutting frame to be more satisfactory. With the universal the cutting direction extends through ninety degrees, and the cutting direction can be finely adjusted for accuracy.

Patterns for the Vertical Cutting Frame

Many patterns can be created with the vertical frame; figure 34 illustrates eight such patterns, all of which should be practiced by the student. The variance in design is accomplished by:

1. The size and contour of the cutter.
2. The length of the segment cut or the number of cuts around the cylinder.
3. The progressive arrangement of the cuts in the successive rows.
4. The combination of different cutters for the same pattern.

BASKETWEAVE PATTERN (fig. 34*a*). In this pattern a square-ended cutter is used to cut sixteen segments around the cylinder. The positions of the segment cuts alternate in every other row so that the cuts fall halfway between each other. A finer pattern can be made by cutting more than sixteen segments around the cylinder.

CHEVRON PATTERN (fig. 34*b*). This pattern also requires the use of the square-ended cutter to cut sixteen segment cuts around the cylinder, but the positions of the successive rows of cuts are different from those in the basketweave pattern. With each successive row of cuts, the index pin is advanced one additional hole past the position of the preceding cut: 6, 12, 18; 7, 13, 19; 8, 14, 20; and so on. This progression continues halfway through the chevron pattern and then is reversed to return the cuts to the starting pin position. This cycle can be a single one or may be repeated several times, depending on the desired design.

SPIRAL PATTERN (fig. 34 *c–e*). This pattern requires the pin to be advanced by one hole for each successive row of cuts through the entire design. A square-ended cutter is used.

BAMBOO PATTERN (fig. 34*d*). In this pattern a concave cutter is used. The index pin is advanced by one hole in the reverse direction for each successive row (similar to cutting the second half of the chevron pattern).

WAVE PATTERN (fig. 34*f–g*). This fascinating pattern is somewhat akin in appearance to rose engine turning and can be used with great effect to decorate surfaces. It makes use of two cutters, the right-hand and left-hand quarter hollow tools shown in figure 27*b*, except that the cutters should be ground to points on the sides and not have the ends square as shown.

The first series of cuts is made with the left-hand cutter, advancing the tool into the work until the trailing edge of the cutter just barely penetrates the surface. The remainder of the first series of cuts is made as already described, with the extreme ends of the cuts just touching. Then the left-hand cutter is replaced by the right-hand cutter. The cuts are repeated, with the intermediate indexing as described for the basket-

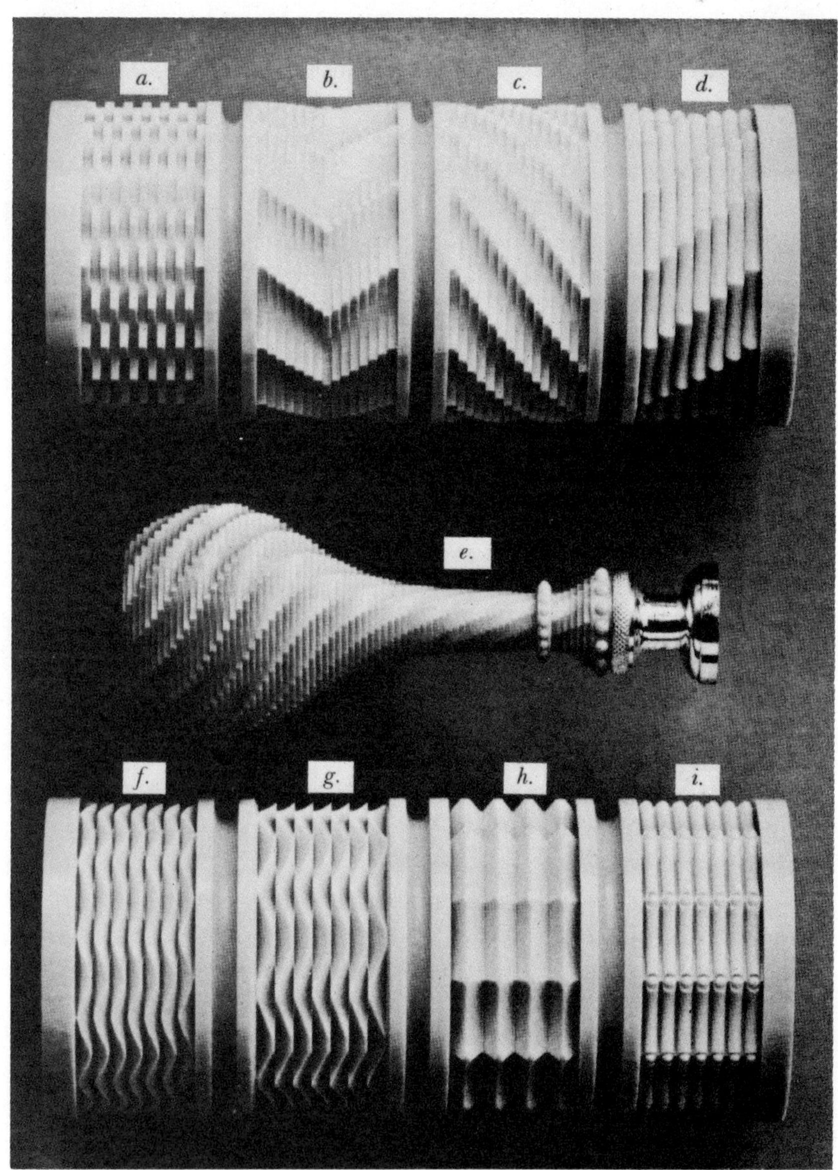

Fig. 34. Typical examples of the application of the vertical cutting frame. (*Reproduced from Plate III*, Holtzapffel, vol. V)

weave pattern. In this pattern it is particularly important that the cutter revolve in a truly vertical plane; otherwise the successive cuts will impinge on the adjacent cuts with visible and negative effect.

ADDITIONAL PATTERNS (fig. 34*g–h*). Other patterns should present no particular problem for the student who has mastered all of the foregoing cuts. The pattern in figure 34*i*, introduces a new tool, the drilling instrument, used to cut a pearl at the junction of each pair of adjacent cuts as for the bamboo pattern. This effect is particularly attractive when arranged in a spiral pattern.

Curved or Contoured Surfaces

The foregoing exercises have been based on horizontal or straight-surfaced cylinders, but the several patterns thus achieved are much more impressive when executed on curved surfaces, such as the ivory stamp handle shown in figure 34*e*. Such patterns are similar to those cut on level cylinders except that the length of the segment cuts must be adjusted to follow the increasing or decreasing circumference of the work. As the vertical cuts proceed across the length of a curve, there is a gradual and regular diminution or enlargement of the length of the segment cuts corresponding to the reduced or enlarged circumference of the work. The greater the curve of the work, the more marked is this change.

The curvilinear apparatus is used to adjust the lengths of the segment cuts to follow the cylinder's curve. This apparatus consists of a bridge mounted on the slide rest above the tool box that holds templates of various contours that conform to the shapes or contours of the work to be decorated (see fig. 23). Mounted in the tool box is a device with a projecting point, a rubber, which rubs against the edge of the template and guides the cutter in a path corresponding to the shape of the template. This apparatus functions like the depth screw in that it controls the depth of cut at any place on the work, but it differs from the depth screw in being infinitely variable in order to follow curves.

Holtzappfel would have preferred the templates be made of steel or brass and held on the bridge by small bolts passing through both template and bridge. No doubt this is practical and desirable if work of the same shape and size is to be done repetitively. Most craftsmen, however, do not mass-produce their work and seldom make two pieces with the exact same shape and size. A template of plastic is therefore more useful. It can be easily reshaped slightly to meet a new contour, or the same piece of plastic can be used again if the change is moderate. One piece of plastic may be used to make several different templates of slightly varying contours. The template is fastened to the bridge with small finger clamps, which allow for adjustment in position laterally in either direction.

THE DOME CHUCK. The dome chuck is one of the most frequently used chucks for special applications. It is used for positioning work to decorate a hemisphere, an operation not possible with the other chucks normally found with an ornamental lathe. This chuck, which fastens onto the mandrel nose, consists of a heavy brass plate suspended vertically toward the bearers. In its chamfered bearings, this plate holds a second plate that can be moved perpendicular to the first by means of a lead screw in the first plate. In this way the work is held perpendicular to the lathe bed and can be positioned toward or away from the height of center of the lathe.

Plate 12 shows candlesticks in which spheres in the base are constructed with the dome chuck. The top and bottom of the spheres were cut separately. Each was mounted on a faceplate with double-faced tape and plain-turned to a hemisphere. It was then mounted in then dome chuck with a vertical cutter mounted in the tool box. Height of center is important and is achieved by moving the piece up and down by means of the lead screw in the dome chuck until the cutter exactly touches both the top center and the diameter of the hemisphere, swinging the chuck through a ninety-degree arc.

With these two points fixed, the revolving cutter is brought to the face of the work. The dome chuck is moved by hand through ninety degrees against the cutter to put the flute on the work. The next flute is positioned with the index plate on the dome chuck, and the work is then fluted to completion.

The Horizontal Cutting Frame

Unlike the vertical cutting frame, the spindle of the horizontal cutting frame stands vertically. It accepts a great variety of cutters, as illustrated in figure 27. Because several different designs of the horizontal frame can be obtained, care must be taken that the throat is large enough to accept the length of the tool being used.

THE HEIGHT OF CENTER. With the horizontal cutting frame, it is most important that the center point of the cutter coincide exactly with the height of center of the work being cut. Height of center is important in all ornamental turnery, and there are several methods by which it is determined.

The elevating screw in the slide rest is used

Plate 1. Covered bowl in three-legged stand made from Brazilian kingwood, African blackwood, African satinwood, pink ivory wood, and Queensland walnut, with a finial of natural ivory. Designed and crafted by the author. (*Collection of Richard Miller*)

Plate 2. Covered bowl with handles in bloodwood, fancy bubinga sapwood, and African blackwood. Designed and crafted by the author. (*Collection of Isabel H. Ault*)

Plate 3. (*Left.*) Chalice in caviuna (rosewood family), purpleheart, Swiss pearwood, African blackwood, and ivory. Designed and crafted by the author.

Plate 4. (*Right.*) A pair of candlesticks in Brazilian tulipwood with Corian (mock ivory). Designed and crafted by the author.

Plate 5. Rare modern work in ornamental turnery from a Holtzapffel lathe. Exquisitely done in ivory and 22-carat granulated gold; the top of the upper piece contains a potpourri chamber. Designed and crafted by Delphin Broussailles.

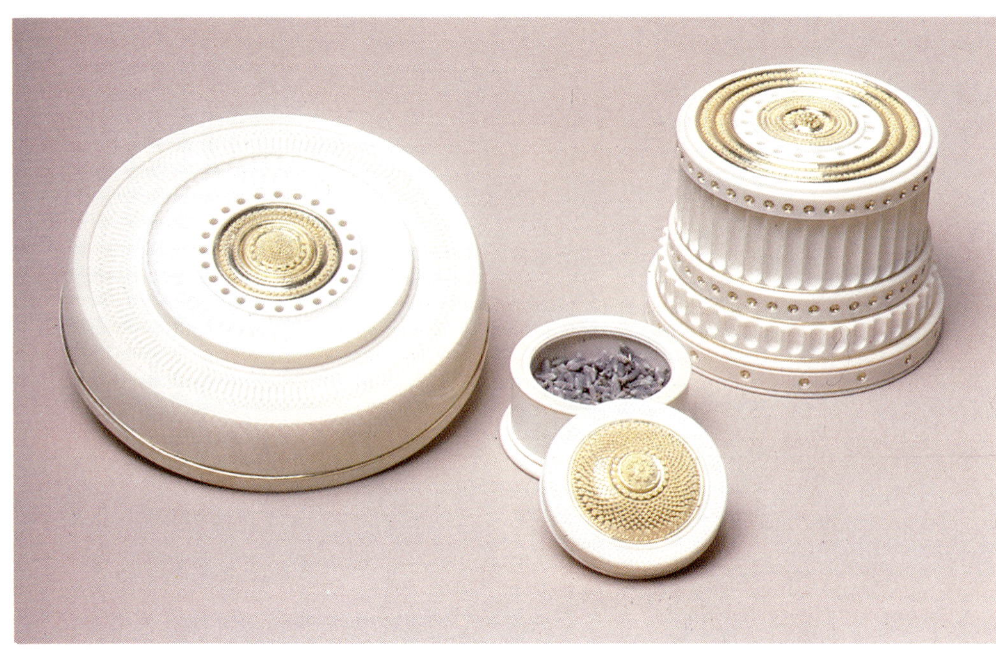

Plate 6. (*Left.*) Candelabrum with twisted spirals. The following woods were used: base, bush willow from South Africa; domes, pink ivory wood from Zululand; stems, apricot wood; cups, holly. Designed and crafted by the author.

Plate 7. Covered box of olive wood. The side cut is the wave pattern (shown in fig. 34, pattern *f*), arranged in a spiral pattern. The cover is a scallop pattern cut with the eccentric cutting frame. The crown was cut with the horizontal cutting frame, and the finial was done with the spiral and curvilinear apparatuses in combination. Designed and crafted by the author. (*Collection of Mr. and Mrs. M. R. McGowan, Jr.*)

Plate 9. Fruit bowl of rare Australian silkwood cut from the stump of the tree with an unusual "mare's-tail" pattern in the grain. The center panel is of ivory with a rim of pearls cut in bloodwood and a center of rare pink ivory from Zululand. The outside is decorated with the wave pattern shown in figure 34, patterns *f* and *g*, but arranged in a spiral pattern. (*Collection of Mr. and Mrs. James B. Woodruff.*)

Plate 8. (*Above.*) Chalice in Brazilian kingwood and ivory. The stem is African blackwood with ivory collars; the lid is edged with the barleycorn pattern in blackwood and capped with an ivory finial and crown. Designed and crafted by the author. (*Photo by Allen Rodney*)

Plate 10. (*Right.*) Formal service plate of imbuia wood from southern Brazil with inlays of pink ivory wood from Zululand. The inlays are decorated with the eccentric cutting frame, showing the varied cutting patterns possible with that instrument. The frieze cuts were made with a drill. Designed and crafted by the author. (*Photo by Allen Rodney*)

Plate 11. (*Above.*) Formal service plate of crotch myrtle wood from the United States, with a center of Carpathian elm burl bordered with pink ivory wood from Zululand. The frieze pattern in the border was cut with the drill instrument. Designed and crafted by the author.

Plate 12. (*Above.*) Fuel-burning candlesticks with wicks in putumju wood (also known as balustre), bloodwood, ebony, pink ivory, and spalted box elder. Designed and crafted by the author. (*Collection of the Mercantile Library, New York, New York*)

Plate 13. (*Left.*) Trophy cup of wamara wood from Guyana. The handles are of pink ivory wood interlaid cross-grain with a darker wood for strength. The finial, center band, and trim are of Du Pont Corian, mock ivory.

Fig. 35. The dome chuck is used to decorate hemispherical shapes. The work is held in a vertical position and can be swung through 90 degrees.

to raise and lower the cutting tool to exact height of center. The screw encircles the post that holds the slide rest in its saddle and provides the means by which the entire slide rest is raised and lowered. The holes around the screw's circumference are for insertion of a pin, which turns the screw. With this elevating screw, the height of center of the cutter can be adjusted to the finest degree.

A simple method for determining when the tool is at the correct height of center is to adjust the slide rest across the bearers and advance the cutting tool until it reaches and aligns with the center of the mandrel nose. If the center of the nose is not easily discernible, a pointed center chuck can be mounted for the purpose. Holtzapffel recommends aligning the cutter with the center point of the work itself; locating the center point on the workpiece is not always easy, especially if it is not a perfect cylinder.

A height-of-center gauge can be made from a tool that was originally intended to determine the accuracy of depth of cut on a circular saw. It is a metal base and frame in which an adjustable post is mounted vertically; a setscrew holds the post in a predetermined vertical position. Drill a small hole horizontally into the tip of the post, and fix a large needle in the hole with epoxy cement. The needle should extend horizontally, parallel to the base. Place the instrument flat on the lathe bearers and raise the post until the point of the needle exactly coincides with the center of the mandrel nose. Tighten the setscrew. This instrument quickly provides a height of center

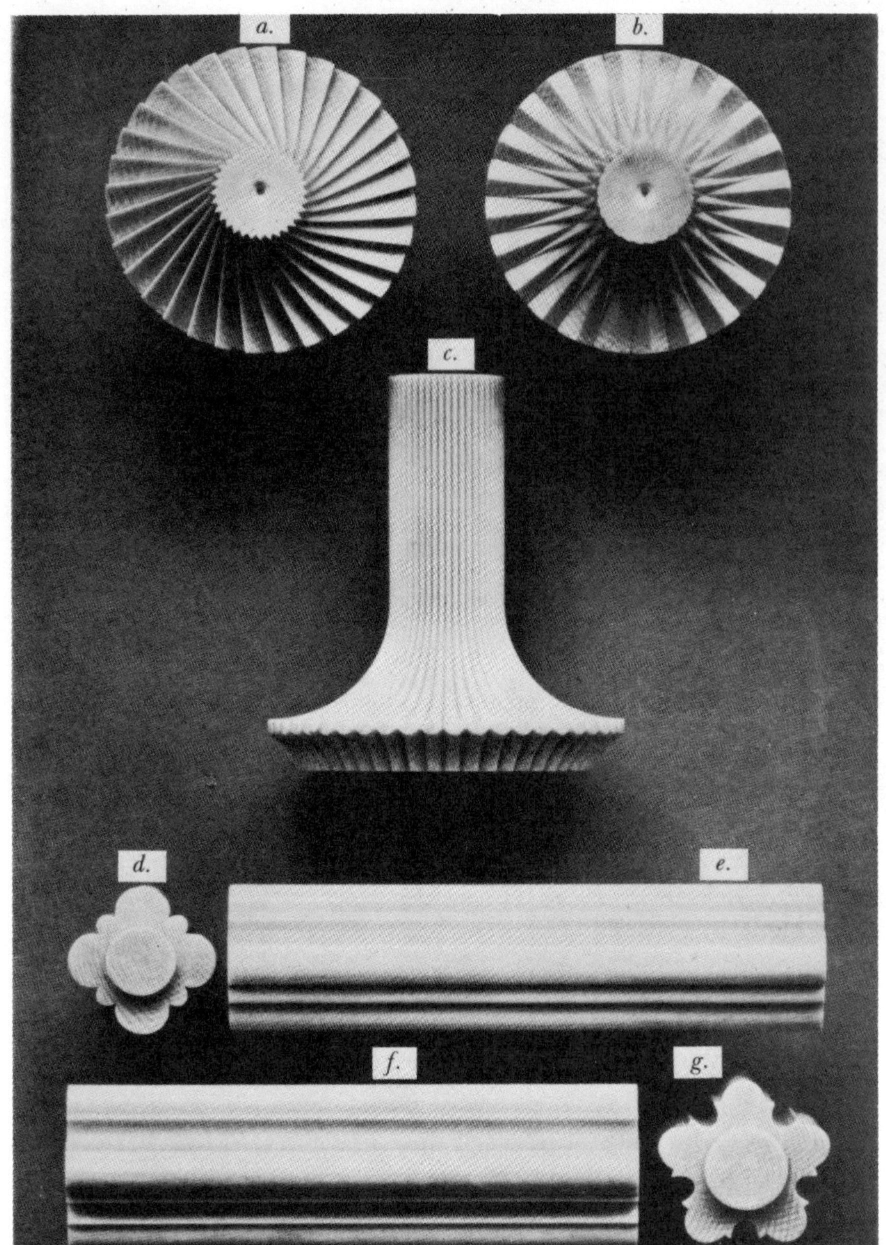

Fig. 36. Reeds or curves and cylinders, clustered columns, and the like, produced with the horizontal and universal cutting frames. (*Reproduced from Plate IV,* Holtzapffel, vol. V)

against which the center point of the cutting tool can be compared.

STRAIGHT-SIDED CYLINDERS. Figure 36 shows some typical patterns cut with the horizontal cutting frame, using a variety of cutters. Figures 36*e–f* are patterns on straight-sided cylinders.

With the selected cutter mounted in the horizontal cutting frame and the exact height of center determined, it is a simple matter to move the cutter horizontally along the cylinder for whatever length is desired, using the lead screw in the slide rest, and cut the desired pattern. In most cases it will be desirable to make a shallow cut first, deepening the cut progressively with subsequent cuts until the desired depth of cut has been achieved. If the length of cut is less than the length of the slide rest, the fluting stops are used. If it is greater than the length of the slide rest, two or more cuts will be necessary, moving the entire slide rest horizontally and matching each subsequent set of cuts with the preceding ones.

CONTOURED SURFACES. If a template were used to guide a cutter along the surface of a straight cylinder, the template would, of course, be straight, to match the surface of the cylinder. A template is normally not necessary with a level cylinder, because the depth gauge can more easily guide the cutter.

When using a horizontal cutter on a contoured surface, a template is necessary to guide the cutter on the curved surface. *At any point on the constantly changing contour of the work surface, the cutter must cut at a depth equal to the radius of the horizontal cutter.*

Figure 37 shows a simple vase with curved surfaces that are to be fluted with the horizontal cutting frame. The first diagram shows the contact made by the cutter held in the cutting frame, with the arc produced having a radius of ¾ inch (*B*). When the cutter is advanced into the work, it will cut a flute to the desired depth. We will assume that we will be making horizontal flutes around the diameter of the vase with the cutter that has a radius of ¾ inch. If the template is made in the exact counterpart of the vase shape at *A*, the result will be a distortion as shown in curve *C*. On the other hand, if the contour of the template is made with its edges exactly ¾ inch from the curvature of the vase at all points, the resultant template shape will be as shown at *D*, and the fluted curve will be made in the original shape of the vase.

The sharper the curvatures of the workpiece, the more obvious will be the discrepancy of the final curve if the template correction is not corrected. The base of the candlestick shown in figure 38 illustrates this point. It shows the increased discrepancy that occurs with more acute curves.

The depth of cut will also affect the contours of the work when the flutes are radiating outward from the center of a disk. This is because the flutes are closer together at the center of the work than at the outer edge, and the apex of the cut will be less than it is at the outer edge. In most cases, however, this difference will be of no consequence and may even enhance the beauty

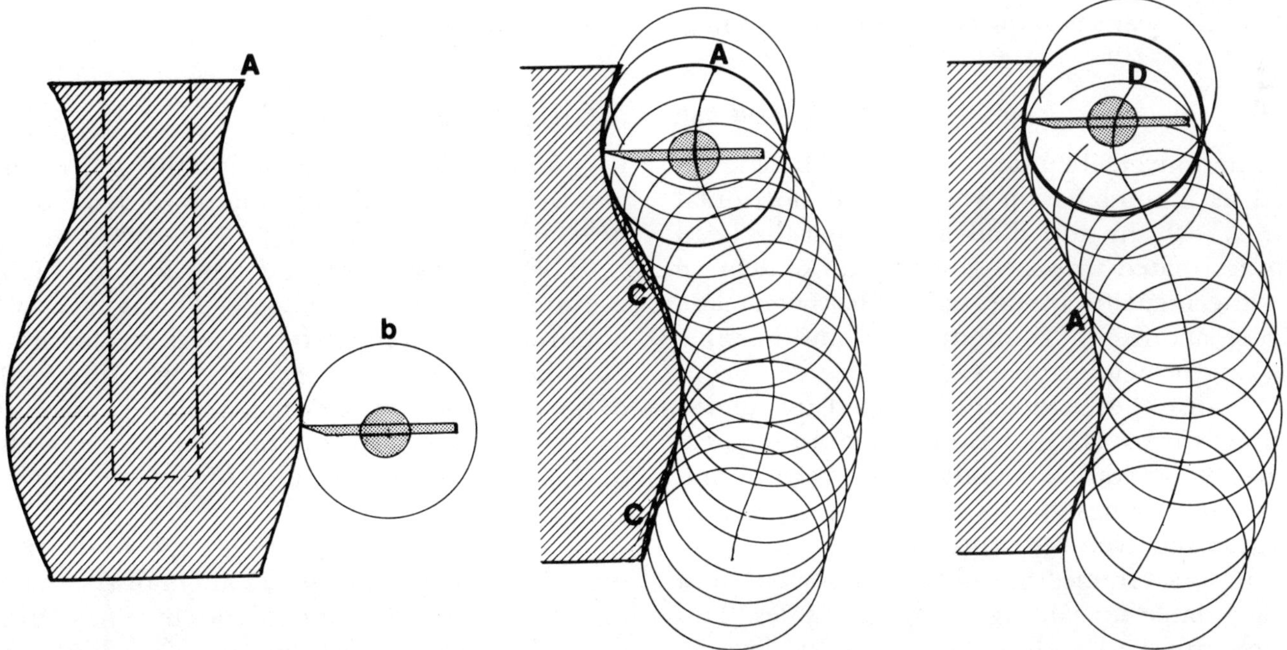

FIG. 37. This diagram shows the method for compensating the template to adjust the radius of the cutter to the proper contour of the workpiece.

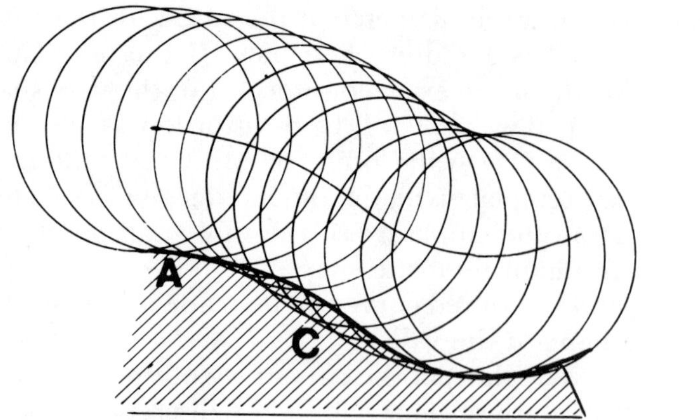

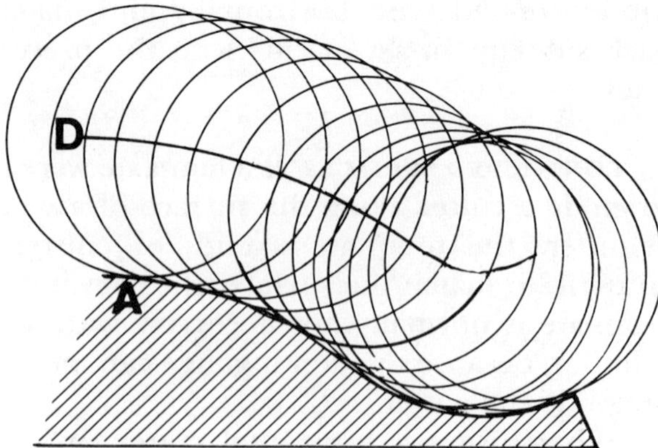

Fig. 38. This diagram shows the technique illustrated in figure 37 applied to a more acute curve in the workpiece.

of the pattern. This is shown on the bases of the candlesticks in plate 12.

A note of caution is in order when the horizontal cutter is being used for spiral or horizontal flutes. In all probability the cutter will be cutting more deeply on one side of the cut than on the other side. After the first cut is made, the cutter will be touching less surface on the side of the cut just made than on the other side, where there is probably some degree of waste wood to be cut away. If the work is being moved clockwise in the index this means that the cutter is exerting more force against the wood on the lower side of the cut than on the upper side. The vibration of the rapidly revolving cutter beating against the lower side of the cuts may well loosen the chuck on the mandrel nose, shifting the workpiece and probably ruining the pattern. The solution is to have the work moving in the index in a counterclockwise direction so that the cutter is beating against the top of the cut instead of the bottom. This tends to tighten the work on the chuck instead of relaxing it and thus eliminates the possibility of loosening or separation.

The Universal Cutting Frame

The universal cutting frame may be considered as a combination vertical and horizontal frame, with the added advantage that the head can be rotated through 180 degrees to achieve any intermediate cutting angle. It has been made in several different configurations, but the same working principle applies to all.

Because the head can be rotated for angled cutting, the overhead drive mechanism requires some adaptation. Whereas, in the vertical and horizontal frames, the belt pulleys are in a fixed position relative to the rotation of the cutter, in the universal frame, the pulley arrangement must be adaptable for cutting at various angles. This is accomplished with a set of three pulleys in combination on the cutting frame. These pulleys permit the head to be rotated to the desired angles without loss of the driving power from overhead.

The universal cutting frame is particularly useful for fluting on an angle. This is especially important with spiral patterns, because the cutter must be arranged in direct relation to the angle of the spiral being cut. After the spiral apparatus has been set up, with the pitch of the spiral determined by the combination of the gear wheels, a tracing cut is made on the work. A pencil line is then drawn at an exact right angle to the tracing cut. This determines the angle of the cutter in relation to the direction of the cut. The cutting frame is rotated so that the cutter is adjusted to that correct position.

The Eccentric Cutting Frame

The eccentric cutting frame is undoubtedly one of the most interesting of all the cutting frames. It consists of a shank that fits in the tool box on the slide rest. The shank is bored lengthwise. A spindle with a T on one end and a drive pulley on the other end is inserted into the shank. This spindle can thus inscribe circles.

ARRANGED CIRCLES. Let us assume that you have a plate mounted on the mandrel and you want to inscribe a pattern of circles on it. Be sure that your workpiece has an absolutely flat surface and that your cutting tool is exactly at the height of center.

In dealing with the eccentric cutter and the groups of circles made by it, two measurements are used to determine the size and position of the group of circles. They are *radius* and *eccentricity*.

Radius. The size of diameter of the circles being cut is determined by the radius of the cutter in the cutting frame, that is, the distance of the point of the cutter from the dead center of the circle being cut. This radius is fixed by the lead screw and the micrometer in the T of the cutting frame. When the cutter is moved to the extreme right of the T head and revolved, it will cut a point. When it is moved to the extreme left of the T head, it will inscribe a circle 3 inches in diameter, with a 1½-inch radius.

Eccentricity. The placement of the group of circles on the plate being decorated will be determined by the eccentricity, the distance of the row of circles from the center of the mandrel or plate. This is determined by the traverse of the lead screw in the main slide rest. Thus, if the point of the cutter is in a position to coincide with the exact center of the plate, there is neither radius nor eccentricity and you will inscribe a point. If the main screw of the slide rest is moved 1⁄10 inch (one revolution of the handle since both lead screws have a pitch of ten threads to the inch), you will still inscribe a point because the radius of the cutter has not been moved. However, if you move the screw in the T head of the cutting frame one revolution, you will inscribe a circle with a 1⁄10-inch radius, ⅕ inch in diameter. If you continue to increase the radius of the cutter but do not change the eccentricity, you will cut a group of concentric circles. In this case the design has no eccentricity, nor has the plate been moved from its original position, which is fixed by the placement of the index pin in a hole in the index head. You have been working with the radius of the cutter but not with the eccentricity of the cutting frame.

Now if you move the cutter out from its center by using the head of its lead screw, to a distance of ½ inch (five turns of the head of the lead screw), and inscribe a circle, you will have a circle of ½-inch radius, 1 inch in diameter. You will still have no eccentricity because you have not moved the lead screw of the slide rest. The edge of your circle will pass through the exact center of the plate. If you proceed to make more circles around the center by moving the index pin a regular number of holes, you will have a figure like that shown in figure 39*b*. Figures 39*a* and *c* are similar, except that in these patterns the index pin was moved to regulate groups of circles that are separated from each other by predetermined space.

Now we introduce eccentricity and move the rows of circles away from the center of the plate. This is done with the lead screw. If the plate is given eccentricity of 1 inch, and the cutter still has a ½-inch radius, you will inscribe a 1-inch circle with its edge ½ inch from the center of the plate. If the index pin is then moved an equal number of spaces, you will create a design similar to that in figure 39*e*. If the pin in the index plate is moved a varying number of holes, you will produce designs like those in figure 39, patterns *d* and *f*.

You can produce four classes of circles using the eccentric cutting frame: centered designs, ring patterns, wide ring patterns, and barleycorn patterns.

Centered designs. Centered designs are created when the eccentricity and the radius are identical. If the eccentricity is 1 inch and the cutter is given a radius of 1 inch, the circles will

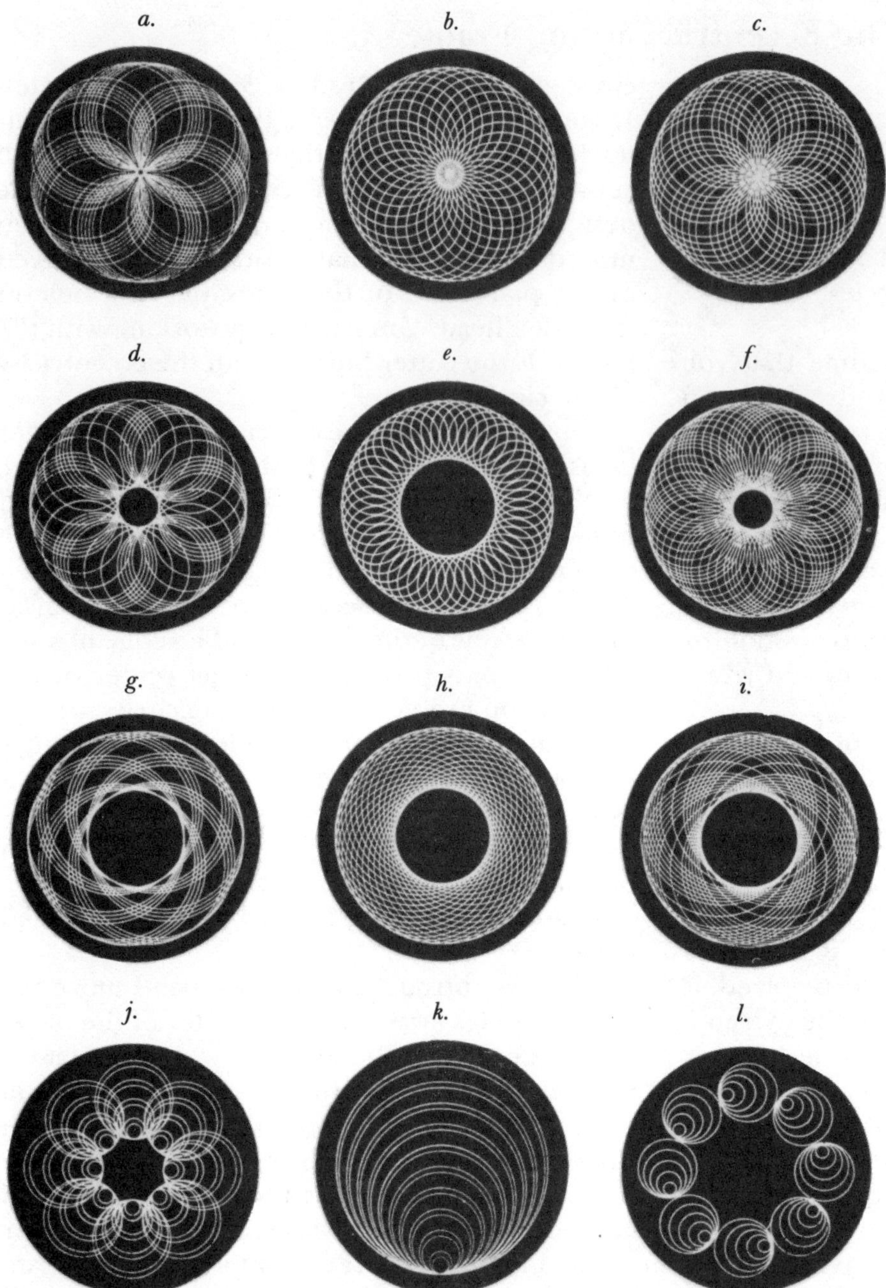

Fig. 39. Typical examples of patterns cut with the eccentric cutting frame. (*Reproduced from Plate IX,* Holtzapffel, vol. V)

be 2 inches in diameter and the edges will pass through the exact center of the figure.

Ring patterns. When eccentricity is greater than radius, the inscribed circles will form a ring in the outer portion of the design. If the eccentricity is 1 inch and the radius is ½ inch, the succession of inscribed circles will form a ring as shown in figure 39*e*. This assumes that the pin in the index was moved the same number of holes each time. By varying the eccentricity and by moving the pin different numbers of spaces, this pattern can be varied to provide a wide range of cutting patterns.

Wide ring patterns. When radius exceeds eccentricity, the result is another type of ring in which the individual circles all pass around to the far side of the center of the pattern. This is shown in figure 39*g–i*.

Fig. 40. This diagram shows the barleycorn pattern when correctly cut, with the edges of the circles exactly touching at their circumferences.

Barleycorn patterns. One of the most attractive patterns produced with the eccentric cutting frame is the barleycorn pattern. This is a deceptively simple pattern; its correct execution requires much careful planning and spacing. It consists of a row of circles placed around the circumference of the workpiece, the circles overlapping each other with the edges exactly touching each other. This leaves small pointed ovals between adjacent circles, which comprise the barleycorns. Because you are essentially creating two patterns at once—the circles and the barleycorns—the radius of the circles and the eccentricity of the row of circles be very carefully determined. (See fig. 39*e*.)

Figure 40 shows the design when correctly spaced, with the circles exactly touching each other at their circumferences. The space between the circumferential junctures produces the central barleycorn pattern, which is further embellished by the curved triangular points around the outer edges of the pattern. Normally the pattern is cut with a double-angled cutter. To make the barleycorns come to an apex, it is necessary to coordinate the spacing with the width of the cutter.

Figure 41 shows the pattern produced when the circumferences of the circles do not touch. In this case either the radius of the cutter or the eccentricity of the cutting frame or both must be adjusted until the circumferences come into contact. If the position of the design on the plate is such that the eccentricity cannot be changed, the correction can be made only with the radius of the cutter; this will either increase or decrease the diameter of the circles.

Figure 42 shows a schematic layout for determining the radius and eccentricity of the barleycorn design. The measurements can be determined using figure 43. The design for figure 43 can be made by mounting a piece of plastic on the lathe mandrel and scribing the radiating lines onto it with a sharp pointed tool, spacing the lines correctly by using the index head. The scribed lines should then be filled in with a color to make them easily visible. With the plastic chart centered on the mandrel, the eccentricity is determined from the work plan. With the barleycorn center at eccentricity, the circle must fall between the lines so that the circumferences exactly touch at top and bottom. Then, by referring to figure 43, you can determine the number of

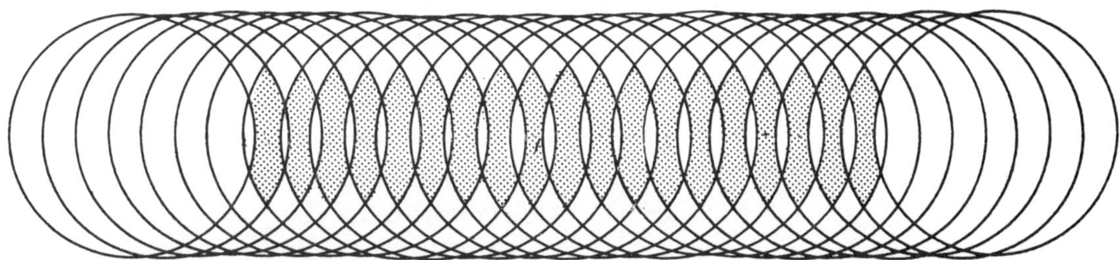

Fig. 41. This diagram shows the barleycorn pattern when incorrectly cut, with the edges of the circles not touching. This "incorrect" cut also produces what may be considered an attractive pattern.

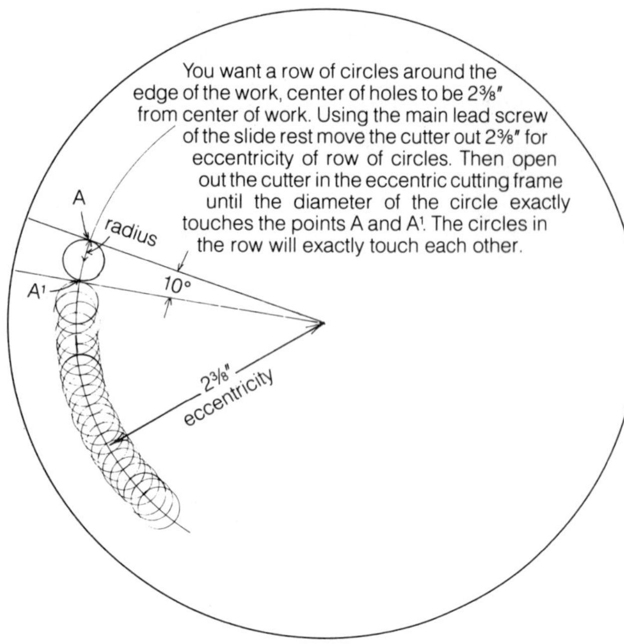

Fig. 42. This diagram shows the relationship of radius and eccentricity for the correct spacing of circles in the barleycorn pattern.

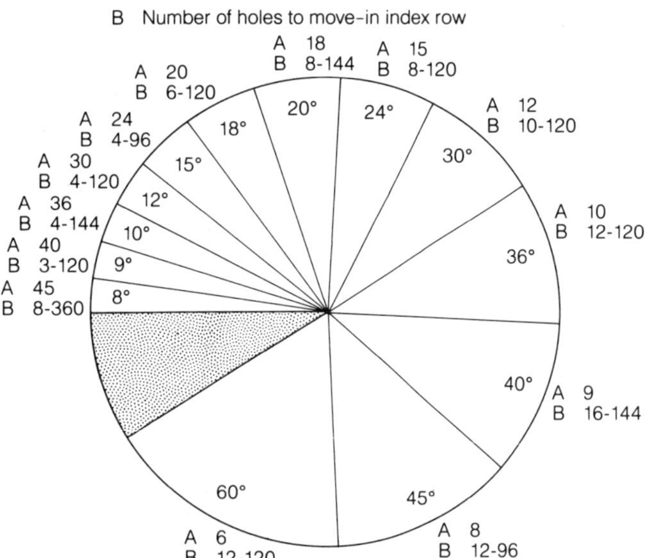

Fig. 43. This chart shows the number of circles to be cut, with the index position for each and the number of degrees of arc.

Index spacing for a given number of divisions of the work

Number of divisions wanted	Move the index pin by the designated number of holes in the following index rows					
	96	**112**	**120**	**144**	**192**	**360**
2	48	56	60	72	96	180
3	32	—	40	48	64	120
4	24	28	30	36	48	90
5	—	—	24	—	—	72
6	16	—	20	24	32	60
7	—	16	—	—	—	—
8	12	14	15	18	24	45
9	—	—	—	—	—	40
10	—	12	—	—	—	36
12	8	—	10	12	16	30
14	—	8	—	—	—	—
15	—	—	8	—	—	24
16	6	7	—	9	12	—
18	—	—	—	8	—	20
20	—	—	6	—	—	18
24	4	—	5	6	8	15
28	—	4	—	—	—	—
30	—	—	4	—	—	12
32	3	—	—	—	6	—
36	—	—	—	4	—	10
40	—	—	3	—	—	9
48	2	—	—	3	4	—
56	—	2	—	—	—	—
60	—	—	2	—	—	6

circles that will complete the row around the plate.

For example, you find that a 15-degree arc will accommodate the circle with the lines just touching at the tops and bottoms. You see from figure 43 that a 15-degree arc, using the 96 row of holes in the indexing plate, will produce twenty-four circles. But you need more than twenty-four circles to create your pattern, so you move the index pin one, two, or three holes to create intermediate circles. This should give you a workable number of circles for your design.

But suppose that the circle did not fit within the arc. Then you have to enlarge or reduce the radius of the cutter to adjust the diameters of the circles to make it fit, or you must change the eccentricity of the row of circles from the center of the pattern. When you have fixed the radius and eccentricity and know the number of holes you must move the index pin to get exactly twenty-four circles, you can then determine how many intermediate holes you want to complete your pattern. Since the twenty-four holes did not fill the pattern, you can cut intermediate circles to fill it out. But, if not the twenty-four circles, then how many?

The table on p. 56 tells you that you can get twenty-four circles with either the 96, 120, 144, 192, or the 360 row of holes in the index plate and the number of holes in the index plate that you should skip to get them. The table tells you which row of index holes to use and how many holes to skip to get your desired basic number of circles, regardless of the number of circles you need.

It may turn out that a trial-and-error method of determining radius and eccentricity will serve if the placement of the cutting pattern on the design is not rigidly indicated. To do this, set up the eccentric cutter with a double-angled cutter and position it in place visually. Move the cutter up to the work and rotate it backward to scribe a faint line on the work. Move the work by the index pin and scribe another faint line. See if the edges meet or, if not, determine how far apart they are and try again, moving either the radius or the eccentricity a small amount to scribe another set of lines. A few tries at this should bring the circumferences into exact position. Then try the spacing of the circles by moving the index pin one or two holes and scribing faint lines until you have a suitable spacing of circles. Normally this will work but not with the exactitude of the method shown in figures 42 and 43.

GRAINING. Graining enhances the reflection of light from the wood's surface. Hard, heavy, dense-grained woods, such as African blackwood, reflects light to a high degree. Cuttings in these woods increase their reflectivity, just as cutting facets into a diamond increases its reflectivity. A sharp tool will produce great beauty in the cuts. In order to achieve the greatest degree of light reflection, the cutter must be razor sharp and the cuts must be made cleanly.

Graining is done by covering the entire surface of the wood with very fine concentric circles cut very close together. Metals are commonly grained by jewelers, as metals have a very fine grain. Blackwood or pink ivory or any other wood with a comparable hard surface can be grained with fifty or even more lines to the inch. One hundred lines to the inch are usually too close to hold as a pattern on wood.

It is important that the depth and spacing of graining circles be precisely determined. This can be done by using the measurement scales on the head of the control screw of the tool box and on the head of the lead screw in the slide rest. Both will give results up to one-hundredth of an inch.

The Drill

This cutting frame consists of a square shank that is drilled longitudinally to accept a spindle with a pulley on one end. A receptacle at the other end of the spindle holds the drill cutters. The aperture for the cutter has a notch at one side; the cutters have a corresponding notch; these hold the cutter in position and keep it from twisting.

The drill is exactly what it seems: it turns on center and cuts a hole. The key to its creative use is found in the drill tools and the patterns they cut. The drill tools illustrated in figure 27 fall into two classes: those in which the pattern is cut central to the axis of revolution of the drill and those in which the pattern is cut to the side of the axis of the revolution.

The contour of the hole cut conforms to the shape of the cutting edge of the drill. Those tools that cut on center can form round, flat, concave, or convex holes. The tools that cut eccentrically form flat, quarter-hollow, concave, or convex holes.

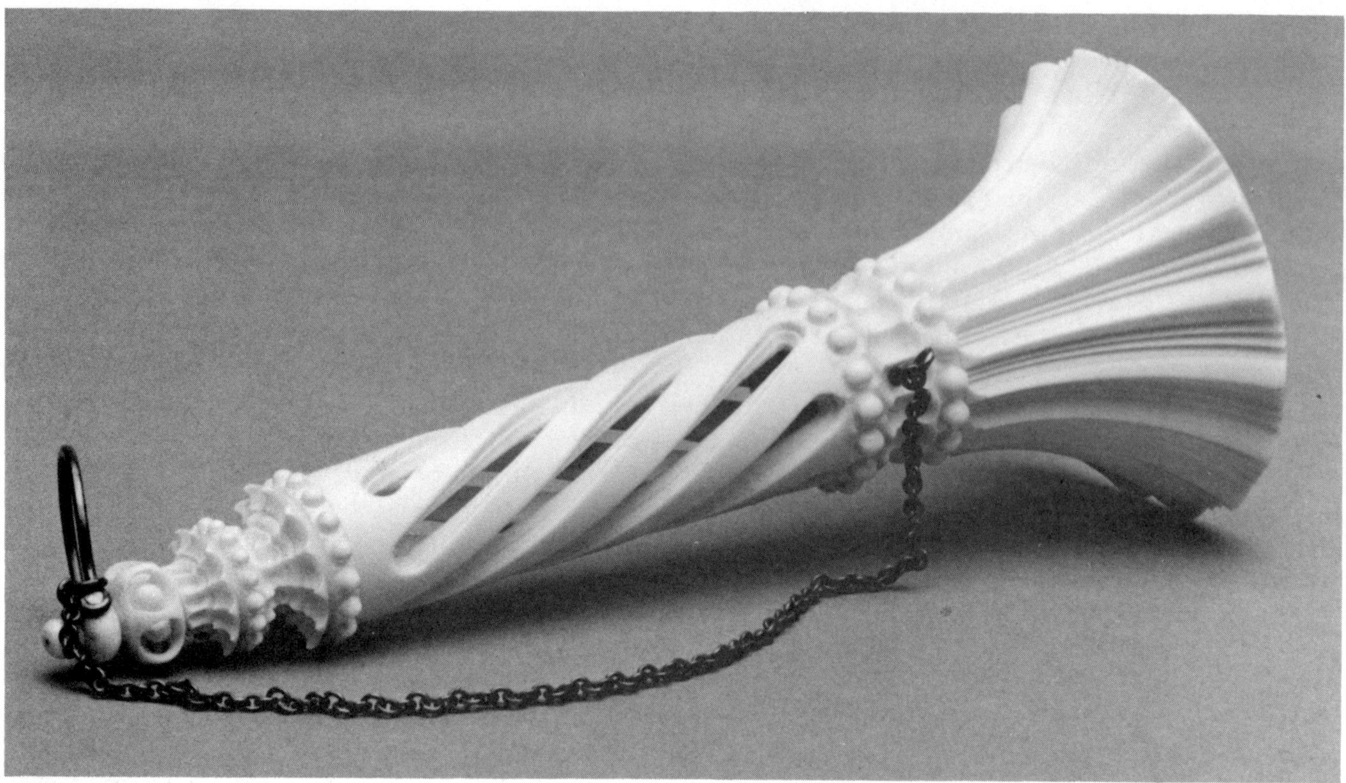

Fig. 44. Antique posey holder in ivory, probably done during the Renaissance, showing a spiral cut made by using the drill-cutting frame holding a cutter with a stairstep pattern. Fluting on the cup was done with a horizontal cutting frame and the curvilinear attachment. The tip shows a complicated pattern created with the drill cutters. The top and bottom are threaded and screwed together. (*Collection of the author*)

The drill, although simple, can produce many complex and beautiful patterns. An excellent application of the drill is seen in the work of Delphin Broussailles in plate 5.

INCISED PATTERNS. The drill can be used quite effectively like a router to cut lines or grooves to form your pattern. Plate 8 shows this application in the ivory band on the chalice. The pattern was laid out in pencil with very careful spacing. The cuts were then made for the radiating lines in the pattern by using the fluting stops to control their lengths. Then the cutter was given eccentricity for the first set of circumferential cuts, the length of the cut being governed by the segment plate on the rear of the headstock. This plate has ninety-six holes with pins to fit in the holes that serve as stops. With this device the circumferential lines were cut accurately to meet the ends of the radial lines previously cut.

Plate 10 shows the use of the drill to cut segments of a frieze pattern interspersed between inlays of finely decorated pink ivory wood. In this case the location and spacing of the nine inlays were determined, and the frieze was laid out in pencil, using the index to space the lines between the inlays properly. The radiating lines were cut by inserting the drill into the wood and using the slide rest lead screw to move the cutter between fluting stops. The circumferential lines were cut by inserting the cutter into the wood and moving the work by hand to interlock the circumferential lines with the radiating lines.

PEARLS. Pearls are cut with a concave drill cutter turning on center with the revolution of the tool. If ivory is being worked, as in the stem of plate 8, there is little problem in cutting pearls. If you are working with wood, however, you may have severe problems because of the grain of the wood. The cutter, in its revolution on center, cuts both with and across the grain of the wood for each pearl. If the wood is not very dense and lacks a strong grain, the cutter may split the wood off when it turns to cut the long grain side. Most woods cannot have pearls cut into them; African blackwood and pink ivory wood can be cut for pearls if you are very careful. Plate 2 shows pearls cut in a wood with grain sufficiently dense to hold the cut without splitting.

STAIRSTEP PATTERN. One of the most interesting cuts made with the drill instrument is with the cutters shaped in a stairstep pattern. These cut on a center and form a groove or flute that resembles stairsteps as seen in figure 44. Many fascinating patterns may be made with this tool, which is shown in the foreground of figure 26.

EIGHT

A Practical Application: Fuel-Burning Candlesticks

TURNING the fuel-burning candlesticks shown in plate 12 involved a number of problems normally encountered in ornamental turnery. Each stick is composed of ten separate pieces. Obviously not all, or even most, of the problems in this project are encountered in the average piece of work. Nevertheless the problems encountered in this piece provide a basis for discussing how to avoid or correct many of the problems inherent in the craft.

The Centerline

The centerline is important in all ornamental turnery, particularly so in the case of tall, assembled pieces, which can instantly be perceived as out of line by the viewer. The centerline should be established at the outset of the work, and all elements should be keyed to it as work progresses. This is illustrated in figure 45.

The centerline is established by drilling the starting hole for the dowel, figure 45*J*, on which the entire assembly is done. This hole must be exactly centerd in the base and at exact right angle to it since it establishes the vertical line of the finished piece. This is accomplished on the ornamental lathe using the tailstock accessories and procedures discussed in chapter 6.

The self-centering 3-jaw chuck is mounted on the tailstock arbor, and a 5⁄16-inch brad-point drill is locked into the chuck. This type of drill is used because it has a very sharp center point that permits it to be brought up to the exact center of the base mounted on the lathe mandrel. With the tailstock locked into position and with the work turning on the lathe at a slow speed of about 700 to 750 r.p.m., the bit is advanced into the base to the proper depth and the control hole is precisely bored. This may seem to be the long way around to the cabinetmaker who has a floor-model power drill, but in actuality, drilling the hole in the absolute center of the base is more easily and quickly done by the above method.

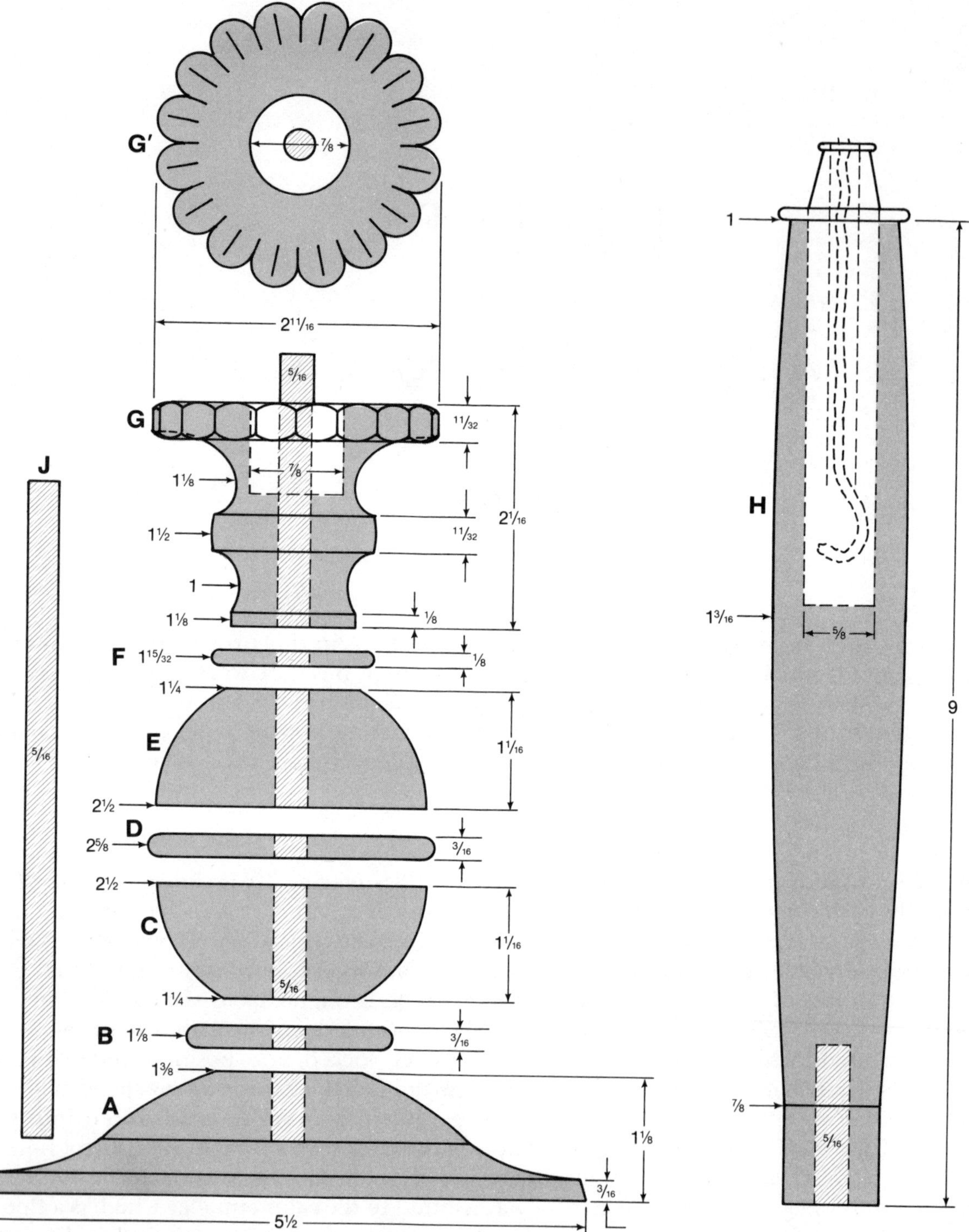

Fig. 45. Schematic working drawing for the fuel-burning candlesticks.

Fluting the Bases

The bases are fluted with a round-nose cutter held either in the horizontal or in the universal cutting frame, discussed in chapter 7. This introduces the problem of template design discussed in that chapter (see *The Horizontal Cutting Frame: Contoured Surfaces*). Figures 37 and 38 illustrate the solution to this problem, with figure 38 applying specifically to the acute curvature of this base.

An alternate correction to the contour of the template may be had by making the template in exact conformance to the contour of the base and then mounting a hemispherical piece on the end of the control screw shown in the center of figure 23. This hemisphere can be made of hardwood (I use lignum vitae) and must be *of the same radius as the radius of revolution of the cutter* in its operation. Since the two radii coincide with each other, the flutes will be in the same contour as the template and no correction is needed. It should be noted that in fluting the base of the candlestick and other such plate designs, the slide rest will be at a right angle to the lathe bearers and not parallel to them as is shown in figure 23, which shows the operation as it applies to the spiral apparatus.

The number of flutes to be cut is a matter of design but is influenced by the number of flutes to be cut on the matching stem sphere since there will be a visual relationship where the flutes meet the collar, as shown in figure 45*B*. This brings into consideration the matter of the flexibility of choice in the available number of flutes. It has been pointed out that the indexing plate shown in figure 18 has a great many pin positions, thus affording a wide variety of choice in the number of flutes. When the spiral apparatus or the dome chuck is used, however, the indexing mechanism is integral to the attachment and in both cases is limited to ninety-six positions. Hence we can make forty-eight flutes on the hemisphere on the dome chuck versus the fifty-six that we can make on the base with the lathe's indexing plate. Without a micrometer, the average observer will not be conscious of this difference. The fifty-six flutes on the base were made by using the 112 row of holes on the indexing plate and moving the pin two holes for each cut.

The cutter must of course be sharpened to a razor edge because there ordinarily is no chance of sanding or otherwise smoothing the finished cuts. Try to finish all the flutes without having to stop and resharpen the cutter, as you may have difficulty in repositioning the cutter exactly.

The Direction of Cutter Revolution

The cutter can revolve to cut outward toward the edge of the plate or inward toward its center. Factors bearing on this decision include the nature of the wood being used and its reaction to the force of the cut. Softer or less dense woods might splinter when the cutter passes across the grain going off the end of the piece if the cut were made outward. In the case of this base, the cutter was revolving to cut inward toward the center, thus passing onto instead of off the edge of the wood, avoiding the possibility of splintering.

In addition to the direction of cutter revolution, the direction of the lateral movement of the cutting frame on the slide rest is important. If working on a straight or level surface, it makes no difference which way it moves. In the case of a piece with acute curves, such as this base, however, the cutter must be moved to cut "downhill" on the template rather than "uphill," to prevent excessive friction of the guidance pin against the upward contour of the template. Contour cuts ordinarily must be made following the downhill shape on the template.

Making the Hemispheres

The spherical design on the stems of these candlesticks required compromise in cutting inasmuch as it is normally not possible to cut a complete sphere on a single piece and then decorate it with flutes. As the project progressed, it became apparent that making two hemispheres and matching them on an intermediate ring or collar not only would solve the problems of chucking and fluting but would actually enhance the final design. The end result bore out this conclusion.

The wood for each hemisphere was mounted on a chuck with double-faced tape after having been perfectly surfaced on one side. The hemisphere was hand-turned to approximately the correct size. The eccentric cutter was then mounted in the slide rest, which was turned to a 45-degree angle to the workpiece. The cutter was advanced to contact the wood and then given eccentricity so that it exactly touched the apex point of the hemisphere at one end of the revolution and the diametric point of the hemisphere base at the other end of the revolution. Thus, the cutter revolution encompassed 90 degrees of the hemisphere with each revolution.

With the cutter driven by the overhead and the mandrel disconnected from its motor, the work was slowly turned by hand to cut an exact hemisphere.

The workpiece was then mounted on the dome chuck (fig. 35) that was vertical to the lathe bearers. Another, slightly smaller round-nose cutter was selected to cut the flutes on the hemisphere and installed in the vertical cutting frame. The cutter was carefully positioned at the exact height of center (see *The Horizontal Cutting Frame: Height of Center* in chapter 7), exactly in line with the diameter of the hemisphere base in order that the flutes would meet at the exact center of the top of the hemisphere. Then the fluting stops (shown at the ends of the slide rest in fig. 10) were brought up and locked in position to prevent any lateral movement of the cutting frame during the operation.

Using the depth stop, the mandrel was rotated by hand through 45 degrees from the outer edge to the apex and was advanced with the control screw until the desired depth of cut was found. The cutter was withdrawn, and the chuck was rotated with the index handle to point zero, using that point as the referent for all further repositioning of the chuck. The flutes were then cut in proper rotation.

LOOSENING OF THE CHUCK ON THE MANDREL. Over the years I have had no fewer than six pieces of work ruined because the chuck worked loose from the mandrel during the cutting process. How does this happen? The chuck is screwed onto the mandrel, seating against the headstock. Once it is seated, the chuck is locked in place by friction only. If this friction fails, the chuck becomes loose and the piece will probably be ruined.

When the cutter cuts a flute, it cuts a tiny channel in the wood. In the process the cutter is "hammering" its way through the wood, even though it is razor sharp and producing a clean cut. Every time the cutter revolves, it strikes the wood with some degree of force. In making the first flute, the cutter strikes the wood equally on both sides of the cut. On the second and each subsequent cut, the cutter is slicing into waste wood on one edge of the flute and against an already clean flute on the other edge. Therefore, a slightly greater force is expended against the waste-wood edge. Since the cutter may revolve one to two thousand times per minute and each flute requires a minute or more of cutting, we are looking at thousands of times the cutter strikes the wood. The build-up of the force of these thrusts of the cutter into the wood is greater on the waste-wood side than on the clean side of the flute. This can cause enough vibration to loosen the chuck on the mandrel and—presto—the cutter bores into already cut flutes, ruining the design. How can that be avoided?

I have talked to several competent toolmakers about devising some way to lock the chuck securely onto the mandrel, but so far I have drawn a blank. Recall that each flute requires the work on the chuck to be in a different position relative to the headstock than in the previous flute. The chuck is screwed onto the mandrel in a clockwise direction. If the work is also being indexed in a clockwise direction, the accumulated force of the thrusts will be against the waste-wood side. Therefore, the answer is to index the piece in a counterclockwise direction, placing the waste wood on the upper rather than on the lower side of the cut. This tends to force the chuck more tightly onto the mandrel, rather than loosening it.

The other solution is to stop the work every few minutes and check to be sure the chuck is tight. This is a nuisance. Counterclockwise indexing is the practical answer.

The Collars

The collars (fig. 45, *B*, *D*, *F*) present no particular problem in turning, although they are related to one that may come up later, in final assembly. There will be six glued joints in the stem assembly, and they must be tight without having glue squeeze out at the ends of the flutes. To accomplish this, the collars, each of which is adjacent to the ends of flutes in adjoining pieces, must have absolutely tight joints before the glue is applied. One way to do this is to surface each collar on each side at the time the center holes are drilled. On the final pass of the cutter, start exactly at the outer edge and then, as the cutter is drawn across the collar with the slide rest handle, advance the cutter into the wood at a gradual slope of perhaps one- or two-thousandths of an inch, producing an almost imperceptable dish shape at the surface of the collar's center. This tiny slope provides a little more space for the glue at the center than at the outer edge. Apply the glue sparingly (never use too much glue) in the central two-thirds of the collar, and let the clamping pressure force the glue outward. No glue should be expelled into the ends of the flutes. If by chance a wee bit should do so, a wet rag at hand can quickly and easily wipe it off before it sets.

Before gluing up, the pieces are assembled and held up against the light. If light shows through between any of the pieces, they are not tight enough, and the preceding technique should be used. If necessary, a small sandpaper disk in an electric drill can be used to make the slope.

Making the Cups

The cups (fig. 45*G*) are begun by plain-turning the stock, which is mounted on faceplates that are at right angles to the bearers. The center hole for the dowel is bored into the cups with the tailstock drilling accessory, and the ⅞-inch hole for the "candle" is bored with a flat-bottom bit to the required depth. The tailstock center support should be used when boring because of the fragility of the wood and the amount of work to be done on it.

Blank wood is left at the tailstock end for the petals. Cutting the petals may present particular problems, depending on the nature of the wood used. I used a spalted box elder, which is very fragile and requires great care in cutting. The sixteen petals are spaced out for position, and the interstices are cut with a pair of quarter-hollow bits mounted in the horizontal cutting frame. Each cutter or bit (fig. 27*b*, third and fourth cutters) produces one side of the petal as it is drawn back and forth by the slide rest handle. Advance very gently into the wood, taking cuts of only a few thousandths of an inch with each pass. When all petals have been cut to the proper depth on one side, the cutters are changed and the second cutter forms the other side of the petal. The groove in the top of each petal is cut with a double-concave bit, with the slide rest repositioned to place the cut in the proper place.

Making the Candlesticks

The holes must be drilled in both ends of the candlestick workpiece (fig. 45). One end will receive the dowel, and the other will hold the glass tube with the wick container. The (dowel) end should have some waste wood, as that end will be held in the chuck. When boring the wick hole, remember that the candlestick will sit in the cup, so that the depth of the dowel hole should equal the seat of the cup plus the length that the dowel will extend into the candlestick. The hole for the glass tube will be too large to be supported by the tailstock center, and a plug with a shoulder should be plain-turned and inserted into this hole to accept the tailstock support. The steadyrest illustrated in figure 25 will help support the workpiece when boring these holes.

Fluting the sticks requires both the curvilinear and the spiral apparatuses, as shown in figures 22 through 24. These important attachments are discussed in chapter 4. Forming the template for the contoured surface of the sticks is discussed in chapter 7 and illustrated in figure 37.

The design calls for a relatively slow spiral on the stem, one complete spiral in four inches. This required a spiral setup (fig. 22) using four wheels of 144, 24, 120, and 18 teeth respectively. The 144-tooth wheel was first in the train and was mounted on the spiral chuck while the 18-tooth wheel was mounted on the end of the lead screw in the slide rest. In order to reverse the direction of the spiral on the second stick, an idler cog was inserted on the banjo arm between the wheels.

The design calls for pairs of flutes, alternating wide with narrow. The spiral chuck contains its own indexing wheel with ninety-six positions, and the flutes were cut by using four spaces for the wide flute and two spaces for the narrow. A double-concave cutter was used to shape the sides of the flutes.

The previous discussion of the possibility of the chuck becoming loose on the mandrel applies equally in spiral fluting and care must be taken to have the cutter operating in the correct direction.

Final Assembly

The entire candlestick is assembled on the lathe, with the base still on its chuck. The tailstock is used to exert pressure. This guarantees that the originally planned centerline will hold. Since the end of the dowel protruding from the cup is not sufficient for tailstock support, a plug is cut to fit into the ⅞-inch hole in the cup and the center of the plug is hollowed to fit over the dowel. A small complication might arise if glue squeezes out of the dowel and cup, gluing the plug into the cup. To avoid this, simply coat the plug with wax, and the glue will not stick to it.

Removing the Work from the Base Chuck

This has been discussed in chapter 6, but in this piece extra pressure was exerted to make sure the double-faced tape did its job. Consequently, a few extra drops of the solvent may have to be used, and a little more work with the spatula may be needed to work the base off the chuck. When the stick is set up for viewing, it will be visually perpendicular and, if well done, a beautiful piece of ornamental turnery.

NINE

Materials

THE amateur ornamental turner uses two basic materials, ivory and wood. Ivory was the preferred material for the majority of the turners during the Renaissance. It was plentiful then and is undoubtedly the best material for making the fine cuts and exquisite designs of the craft. Turners such as Zick, Nartov, Treffler, Barreau, and Heiden, as well as royalty, preferred using ivory over all other materials.

Ivory

Today, because the animals that produce ivory are on the endangered list in many countries, ivory is becoming more and more scarce. When it is available, the cost is prohibitive to the average amateur. If and when it is available, it should be used sparingly for trim, to create contrast with such woods as African blackwood. Finials offer a splendid use for ivory.

Fortunately, a synthetic material has recently come on the market that offers the craftsman an ideal substitute for ivory. Called Corian® and made by Du Pont, it was intended for use in architectural and decorative applications, mainly in the kitchen, bathroom, and bar. The innovative craftsman, however, will find many special uses for Corian in his or her work. It is a solid material, nonporous and highly stain resistant. It is available in white and off-white colors, one of which is practically indistinguishable from ivory. The smooth surface is durable and, when polished with tripoli, takes on a beautiful sheen.

The chalice in plate 8 shows the use of ivory, as do the Broussailles work in plate 5 and the pieces in plate 9 and figures 29 and 44, whereas the trophy cup in plate 13 and the candlesticks in plate 4 show a comparable use of Corian. The new material machines extremely well under the cutters in ornamental turnery and, although a mineral-filled substance, does not dull the cutters any more than ivory does. Used in conjunction with beautiful woods, this material offers new challenges to the ornamental turner.

Wood

For the average amateur, wood remains the most available material, even though many of the choice tropical woods are becoming more difficult to obtain. Many of the world's tropical countries have embargoed logs of most of the choicest woods, although they will cut the logs and send the cut pieces. This limits the availability of the choicest woods in log form, which are preferred by most ornamental turners, who work largely in the round rather than the flat.

To accommodate the intricate designs cut in ornamental turnery, the wood must be able to accept and hold the very fine cuts without splitting or splintering. Some woods will accept the cut but leave "whiskers" at the edge of the cut, especially when cut across the grain. Since it is practically impossible to sand the wood after fine cutting, these woods should be avoided. Use only woods with very hard and dense grain patterns.

Woods are commonly classified as soft and hard, although to the wood scientist this provides only the grossest description. Hardwoods come from deciduous trees and softwoods come from coniferous trees. The characteristic in which the ornamental turner is interested is the density of the grain and its ability to accept and hold the fine cuts. Open-grain woods are to be avoided unless the cutting pattern is to be in the direction of the grain only and not across the grain. The following are some of the woods particularly good for ornamental turnery if and when they can be obtained.

AFRICAN BLACKWOOD *(DALBERGIA MELANOXYLON)*. African blackwood is one of the blackest of woods, although its color sometimes takes on a dark purplish plum cast. It comes from the east coast of Africa and is also known as Mozambique ebony, although true ebony is of the *Diospyros* rather than *Dalbergia* genus. Holtzapffel said, "It is most admirably suited to eccentric turning as the wood is particularly hard, close and free from pores, but not destructive to the tools, from which, when they are in proper condition, it receives a brilliant polish." Alexander Howard, author of *Timbers of the World*, says, "It has a rare quality, namely that unlike practically every other wood, it does not shrink on either way of the grain, retaining exact measurement after machining, so that it is especially useful for pattern making, for screws and the like."

Blackwood has a very narrow, light-colored sapwood. The demarcation between the sapwood and heartwood is one of the sharpest of all woods. The ornamental turner must search for blackwood, since it is seldom available in the open market and is apt to be quite expensive when found. The weight of blackwood is about 90 pounds to the cubic foot, as opposed to mahogany, which weighs about 40 pounds. It is one of the hardest and heaviest of all woods. The tree grows to a diameter of 10 or 12 inches, is slow growing, and is apt to have shakes or splits in its center.

African blackwood has also been known under the names grenadillo, Senegal ebony, and a dozen or more local names in its growing region.

LIGNUM VITAE *(GUAIACUM OFFICINALE)*. Lignum vitae is another of the heaviest of woods, weighing about 89 pounds to the cubic foot. It grows in the West Indies to a size of about 8 to 12 inches in diameter. The heartwood is of a greenish black color, subject to wide variation in pattern, while the sapwood is bright light yellow. It is used for sheaves in blocks-and-tackle, as well as for bearings for propeller shafts on even the biggest of ships. The wood was used for wassail bowls in the seventeenth and eighteenth centuries.

Lignum vitae contains a large content of oil and must be finished carefully, since lacquers will not adhere to it. An oil finish is probably best, although the polyurethanes will take. Just polishing the surface as it comes off the tool is as good a finish as any for most purposes, and it will retain a good polish for long periods. It is commercially available, but supplies are limited because of its use in shipbuilding. It is very good for ornamental turnery, since it will take and hold very fine cuts without splintering or whiskering. It is apt to be hard on tools, which may have to be sharpened during a cutting session. Lignum vitae is capable of producing wide variation in color and grain pattern and is undoubtedly one of the best woods for the ornamental turner.

Lignum vitae is also known in some areas as lignum sanctum, palo santo, and other purely local names. I have used palo santo but found it to be somewhat different in color and grain pattern and lighter in weight than lignum vitae. It works well in ornamental turnery.

BULLETWOOD OR BEEFWOOD. Kribs lists this wood as *Manilkara bidentata*, while Howard lists it as *Humiria floribunda*, both apparently being the same wood. Other names ascribed to it by Kribs include massaranduba, sapodilla, balata, and many others in local areas. Its color ranges from a dull plum-red color (most common) to light or dark red, sometimes with a slightly oily appearance.

I made my wife a cutting board from beefwood some twenty-five years ago, and it has been used constantly in the interim. It shows only slight cut marks on its surface, with no splintering or defects or change in shape despite constant washing.

Beefwood is not customarily seen in even the rare wood markets, but if and when it is available, it is among the costly woods. It works exceedingly well in ornamental turnery, and its rich color makes it a good choice to use in conjunction with other lighter-colored woods.

PURPLEHEART *(PELTOGYNE PANICULATA)*. This wood is especially desirable for ornamental turnery due to the combination of weight and hardness with its beautiful color. It has a dense close texture and takes a beautiful finish. The color may range from a deep plum to bright purple. An unusual characteristic is that the rich purple color holds its brilliance with age and does not fade as do some other tropical woods. It takes and holds fine cuts and will not fuzz or feather under the cutter. It is among the rare woods in the market, although at times it is possible to find the best color. It weighs 66 pounds per cubic foot and comes from Surinam and Guyana. Howard says that the Surinam wood is superior to that from Guyana. If the wood has good color, it will blend with most of the tropical hardwoods in making built-up pieces.

Purpleheart is also known as amaranth, violet wood, bois violet, and by many other local names.

KINGWOOD *(DALBERGIA CEARENSIS)*. Brazilian kingwood is also known as marnut, although Kribs notes that marnut is not a true kingwood but belongs to the genus *Machaerium*. The color and grain pattern of kingwood make it one of the most beautiful and sought after of all turnery woods. Growth rings give it concentric layers of light violet brown and deep purple to almost black. It is very heavy, weighing 75 pounds to the cubic foot, and has a close grain.

A turning made of kingwood should be decorated with much restraint since the grain pattern inherent in the wood normally produces a decorative effect far beyond that given by the turner. Ornamental cutting patterns can easily "fight" with the grain pattern. Used in contrast with other lighter-colored woods in assembled pieces, it can create very beautiful effects (see plate 1).

IRONWOOD *(OLNEYA TESOTA)*. Many of the tropical countries have a wood which they call ironwood, but the botanically true ironwood is found in the southwestern United States. Howard lists fifteen different botanical names of woods, all of which are known locally as ironwood, and many have the weight and hardness to qualify for the name. *Olneya tesota* weighs 66 pounds per cubic foot and is deep chocolate brown to black, mottled with yellowish red. The wood is very hard on cutters, requiring constant sharpening. Very scarce, it is difficult to find and expensive when located.

PINK IVORY *(BERCHEMI ZEHERI)*. One of the world's truly scarce woods, pink ivory is also one of the heaviest and hardest. About 65 pounds to the cubic foot, it is pink to light red in color, with little or no grain pattern. It grows in Zululand in Africa and is considered by the natives to be a sacred tree. Cutting the pink ivory tree, when and if one can be found, is now prohibited without the permission of the South African government. At an auction of the International Wood Collectors Society, a piece of pink ivory measuring 12¾ by 2¼ by 1¾ inches sold for $103, a record price of $2.05 per cubic inch. Some thirty years ago, I obtained a log of pink ivory measuring 5 feet long by about 10 inches in diameter. It took two people to lift the log, and we ruined several saws when cutting it. A few pieces from this log are still the most precious possessions in my wood supply. Needless to say, at that time it was not priced by the cubic inch and also was practically unknown to craftsmen.

BOXWOOD *(BUXUS SEMPERVIRENS)*. Cultivated in English gardens, this tree grows from Europe to the Mediterranean and as far as India and even Japan. In the northern areas, it is more of a shrub, but in the south it grows to tree size. Persian boxwood is considered one of the most desirable for craftwork. The wood is light yellow in color, very dense, with almost uniform texture and very fine grain. It is used for engraving blocks, engineering and mathematical instruments, woodwind musical instruments, and inlay work. Abasion boxwood (close to Turkish) is used for wood engraving and cannot be satisfactorily substituted for, according to Howard. British boxwood weighs about 70 pounds to the cubic foot, while Indian boxwood weighs about 55 to 60 pounds. It is an excellent wood for ornamental turnery, especially as a contrast to some of the darker and more patterned woods.

SATINWOOD *(CHLOROXYLON SWIETENIA)*. Satinwood is a very fine wood that comes from several localities in the Far East; a variation comes from the East Indies, although this is of the *Zanthoxylum* genus. Ceylon is the center of the growing area. It is golden yellow in color and works well as a contrast in inlay work. Its weight is about 50 to 55 pounds per cubic foot. It works well in ornamental turnery, although it is given to surface cracking. It was a favorite with Renaissance craftsmen as an inlay wood. Among fine cabinetmakers satinwood is known for the variety and beauty in its grain patterns, among which are the plain, striped, bee's-wing, and rippled grains. All can be used to good effect, especially in combination with other fine woods, particularly fine mahoganies.

ROSEWOOD. This beautiful wood comes primarily from two areas: Brazilian rosewood is *Dalbergia nigra* and East Indian rosewood is *Dalbergia latifolia*. Both woods weigh in the area of 50 to 55 pounds to the cubic foot and are hard and heavy. Brazilian rosewood is considered to be somewhat more beautiful than East Indian, although both are very attractive and are considered to be among the choicest of cabinet woods. The Brazilian normally has a greater variety of figure in the grain than the East Indian does. They are less valuable in ornamental turnery than in cabinetwork because of a tendency toward surface cracks and open pores.

EBONY. The name *ebony* is used for a variety of woods, some of which are true ebonies and others which are black but not true ebony. All true ebonies belong to one botanical family, Ebonaceae, and to the genus *Diospyros*. Not all ebonies are black, especially Macassar ebony, which has lighter streaks of color, sometimes very marked in color variation. Ebonies are hard and heavy woods ranging from 62 to 75 pounds per cubic foot in weight. For ornamental turnery African blackwood is far superior to ebony, being of a more uniform blackness and density.

OTHER WOODS. Many woods may be used in ornamental turnery other than those mentioned above, but these will indicate the general characteristics to be sought in fine work. The main characteristic is the density of grain necessary to hold the fine cuts made in ornamental patterns. African blackwood and lignum vitae may be considered as benchmarks in this regard. A strong secondary characteristic is the nature of the design being worked and the color and grain pattern necessary to carry out the design. Variations in choice of wood in the same piece will be a strong factor for contrast in both plain and fancy areas of the design. In plate 1 the contrast in grain pattern between the bowl and the base as well as the variation in color are important to the overall design. It would aesthetically kill the piece to have as much grain pattern in the base as appears in the bowl, yet carrying out the color of the bowl in the base ties the two areas together.

The challenge of design and use of color and grain pattern in wood is one of the most interesting factors in ornamental turnery, and the turner can exercise his skill in this regard to his heart's content.

TEN

Ornamental Turnery Today

CREATIVITY is the word that best applies to ornamental turnery. The craft challenges the craftsman to take a plain-turned piece of wood and embellish or ornament it to create a totally new object of art. To choose a piece of rare wood with great natural beauty and embellish it by accentuating the wood's inherent beauty, to create a patterned texture on the surface of the wood to enhance the color and grain pattern, to combine two or more pieces of attractive wood, blending them together through the addition of ornamental decoration—all these possibilities should be made available to more craftsmen.

When the industrial revolution reduced much of the incentive for the individual to create a complete piece of work himself, ornamental turnery was one of the crafts that suffered. Now it is regaining some of its former recognition as one of the techniques that, over the centuries, produced some of the most beautiful, interesting, and sometimes baffling pieces in the creative arts.

Crafts in general are enjoying a tremendous resurgence. Individual creativity and, simultaneously, a greater appreciation of the traditional art-and-craft forms of the Renaissance and Victorian periods are possibly reactions against the extremes of the avant-garde contemporary art forms of the twentieth century. In this resurgence ornamental turnery seeks a place, but the would-be ornamental turner encounters the obstacle of not being able to find the necessary lathe on which to do the work.

Although Holtzapffel made more than 2,500 lathes and others made lesser numbers, they are not to be found today other than by rarest chance or by spending considerable sums of money when one comes on the auction market from old estates or is found in piles of discarded machinery.

Making ornamental turnery available to the modern turner requires that some machine toolmaker produce the relatively small number of attachments to the modern lathe that make ornamental work possible. One large machine-tool company, on being approached with this proposal, immediately countered with the question "How many could we sell?" Not having made a market survey, I was, of course, unable to answer, and they immediately lost interest. But, after practicing the craft for some twenty years, I have had a tremendous amount of correspondence as well as visits to my shop to see the work being done, and the question "Where can I get a lathe?" is always asked.

The success of the modern reprints of volumes 4 and 5 of Holtzapffel's classic *Turning and Mechanical Manipulation* indicates a considerable and growing interest in the craft. Ornamental lathes coming on the market in recent years, few as they are, have sold for many thousands of dollars. Unfortunately, even at those prices, many such lathes come to the buyer with essential parts missing or in a condition requiring major repair work to make them operative, thus adding to the cost.

Giving the modern lathe ornamental capabilities would be no great job for a toolmaker. It is my opinion that the market would justify the effort with considerable profit. What would be necessary?

First, of course, would be a plain lathe, of which there are thousands in use. Then attachments would have to be made to fit the lathe. The first of these is a slide rest such as that shown in figure 10. This slide rest, although similar in function to the slide rest on most metalworking lathes, has some essential differences. First, it is much more flexible and has a lateral capability of 13 inches rather than the 5 or 6 inches common to most metalworking lathes. Second, it is much lighter in weight, which is important because the rest must be removed and then replaced on the lathe, sometimes four or five times in the creation of a single piece.

The tool box on the slide rest must be capable of strict depth control but also be freed for continuous but controlled change of depth in cut for curvilinear work. Some members of the Society of Ornamental Turners have adapted the metal slide to do ornamental work, but it places restrictions not present with the true ornamental slide rest. Effective ornamental turning requires the more flexible slide rest.

Fastening the rest to the lathe bed would require consideration of the variety of shapes and dimensions of the bearers on modern lathes. It is probable, however, that the manufacturers of lathes would want to take advantage of the ornamental market by making saddles to fit their lathe bearers that would accept the standard ornamental rest. This should be no problem.

The second attachment would be the overhead drive, which activates the rotating cutters. The overhead on the Munro lathe shown in figure 12 is the type most commonly encountered, either as shown or adapted in some way. However, the type of overhead drive on my lathe, shown in figure 11 (which is certainly not the original overhead), is, in my opinion, much more efficient and would be easier to construct and to attach to the modern lathe. Making this overhead drive should fall within the capability of many hobbyists with a modicum of industrial arts training.

The next attachment to consider would be the cutting frames, as shown in figure 21, which fit in the tool box on the slide rest and hold the cutters. Only three of these are needed for the turner to cut a wide variety of patterns. The universal holder (fig. 21*A*) would be important, since it serves for both horizontal or vertical cutting patterns as well as angled cutting through 180 degrees of position. The second would be the drill holder. Figure 21*C* shows an original Holtzapffel drill holder, but figure 21*D* shows the modern adaptation, which, with the 3-jaw chuck, is more efficient and is not difficult to make. The eccentric holder cutting frame (fig 21*F*) is an original Holtzapffel model but has been counterbalanced to diminish vibration. This is an important tool holder and is used in circular cutting patterns which are integral to a great many cuts.

The fourth attachment, the indexing plate, is extremely important (see figs. 11 and 18). This is used in a majority of cutting patterns and guarantees the correct spacing of the cuts. On the standard ornamental lathe, it was made integral to the headstock. This poses a problem on the modern lathe, on which the headstock is normally enclosed with no place to attach an indexing plate. However, a solution could be in making a faceplate to go on the mandrel nose, as do all the other chucks, and putting the indexing holes in this faceplate. The faceplate, in turn, would carry an auxiliary mandrel nose to accept another chuck holding the work. This would guarantee concentricity of the index holes, an absolute necessity in ornamental turnery.

This leads to the cutters themselves. Holtzapffel made beautiful mahogany boxes with sloping tops and drawers in which could be placed as many as five or six hundred cutters. The number of cutters with any given lathe varied in the beginning and varied even more with the passage of time. One modern turner obtained a lathe, but when it was delivered, he found exactly one cutter—something like an automobile with a pint of gasoline. On the other hand, a collection of more than 800 cutters came on the market in a recent auction. For the modern turner, 25 or 30 cutters would make possible a good number of cutting patterns, and about 150 would give him an essential range of sizes and shapes. The cutters would present no problem to the modern toolmaker.

Thus, the toolmaker would be faced with making the slide rest, the overhead drive, three cutting frames, the indexing plate or chuck, and the cutters themselves. With modern machine-tool capabilities, this should present no great problem, the only missing link being the manufacturer with vision to see the market.

No one today is making ornamental attachments for the modern lathe. True, some are making attachments to hold a router on the lathe bed to move laterally with a lead screw, making possible flutes and spiral cuts within rather limited ranges. This is probably suitable to the industrial turner who makes chair and stair spindles in quantity. For the amateur turner who is turning as a hobby, this arrangement is quite cumbersome and will not produce the many exotic cuts that constitute ornamental turnery. Some amateurs have adapted the metalworking lathe to do limited ornamental cuts but it is an arduous and time-consuming task with limited results.

It is time for ornamental turnery to come back into its own as a creative and rewarding hobby and take its rightful place in the field of craftsmanship. There are thousands of turners working today, many with very high degrees of skill. There are thousands, even millions, of persons who badly need a creative hobby to fill their time. Isn't it time for some capable company or individual to start the wheels rolling to bring ornamental turnery to these people? The shades of the ornamental turners of the nineteenth century would probably smile and say, "It's about time!"

POSTSCRIPT

Since the completion of the manuscript for this book, a complete prototype of an ornamental lathe in the Holtzapffel manner has been made by the Lawler Gear Company of Raytown, Missouri, for potential marketing internationally (fig. 46). Ray Lawler contacted me with regard to the effort, asking assistance in the definition of measurements for the lathe, Holtzapffel, in volume five of *Turning and Mechanical Manipulation*, carefully made drawings of each and every part of the lathe, but there are no measurements. As a result Mr. Lawler spent two days with me in my shop with a micrometer and camera, taking measurements and pictures of my Holtzapffel lathe.

I have not had an opportunity as of this date to work the Lawler lathe, but Walter Balliet has seen the lathe and reports that it meets all expectations for ornamental turnery. It is to be hoped that this lathe will fill a large gap in the exercise of this rewarding craft and furnish creative craftsmen with the means to fulfill their desire to create.

Fig. 46. A 1985 prototype ornamental lathe made in the Holtzapffel manner by the Lawler Gear Company of Raytown, Missouri, for potential international marketing.

INDEX

Romain Jacquet-Lagrèze

I grew up in the suburbs of Paris – an environment that never offered me much in the way of Hong Kong-related things. However, I remember enjoying some Chow Yun-fat action movies during my childhood, and later on, some films from Wong Kar-wai. I never imagined I would one day be living in this far-off place.

After studying multimedia and art, I went abroad to work in America and Japan. My path finally landed me in Hong Kong in 2009. I settled into the Jordan district of Kowloon and started adapting to my new lifestyle. I quickly fell in love with the authenticity of its streets, the open-air market, and its countless delicious restaurants.

As a graphic designer, I initially felt it more natural to express my vision of this city through drawing. But little by little, I found my camera to be more efficient in capturing the vibrancy of the urban scenery. I started shooting more and more pictures and exploring more and more new districts until I accumulated a satisfying collection of images.

在巴黎近郊長大的我，實在沒有太多機會認識香港，只記得兒時很愛看周潤發的動作片，還有王家衛的電影。想不到長大後，我竟然會來到香港生活，感覺像是機緣巧合，也彷彿是命中註定。

我在大學主修多媒體和藝術，畢業後分別到過美國和日本工作，直到2009年才首次踏足香港，並在九龍的佐敦區展開了我的新生活。從那個時候開始，我已經深深愛上了區內的大街小巷和露天市集，還有無窮無盡的特色美食。

我一直從事平面設計工作，習慣以繪畫表達想法，但漸漸才發現，攝影更能仔細刻劃自己對香港的感覺，所以我嘗試拿起鏡頭，在城市森林中不斷探索，並將最滿意的作品結集成書。

Vertical Horizon
香港印象

Romain Jacquet-Lagrèze
www.rjl-art.com

Design by Romain Jacquet-Lagrèze
Text by Romain Jacquet-Lagrèze & Ashbi Ng
Colour management by Asia One Graphic Limited
Printing by Asia One Printing Limited

asiaone
Published by Asia One Books
aopp@asiaone.com.hk
www.asiaonebooks.com

First published in September 2012

ISBN 978-988-15316-8-1